Permaculture Volume 1

AF880747

Selected articles from the Permaculture Association of Western Australia

Ross Mars and Bob Willis (Editors)

Permaculture Volume 1
The Best of PAWA

Selected Articles from the Newsletters of the Permaculture Association of Western Australia
1978 to 1990

Ross Mars and Bob Willis (Editors)

Originally Published by Candlelight Trust. 4110 Phillips Rd, Mundaring. 6073

Acknowledgements

A sincere thank you goes to the following people who have made valuable contributions to this book:

Foreword: Bill Mollison

The Formation of PAWA: Ros Hart

Proofreading: Pat Scott, Julie Woodman, Margaret Sampey, Alan Thomas Gray

Artwork: Patricia Dundas

Typesetting and layout: Ross Mars

Original Cover: Ian Sergeant, Visarge Photography. Cover changed for this edition.

Copyright © 1996 Candlelight Trust. Reformatted 2014.

Apart from any fair dealing for the purposes of study, research, criticism or review, as permitted under the Copyright Act, no part of this book may be reproduced, stored or transmitted by any process without the written permission of Candlelight Trust, which will act on behalf of the individual authors of these articles and the Permaculture Association of Western Australia.

Warning: The copyright expressly forbids any individual, organisation or school from making additional copies for themselves or other individuals, or for use in other schools or organisations.

Every effort has been made to contact individual authors. However, this has not been completely successful. Enquiries about the contents of this publication can be directed to Candlelight Trust.

National Library of Australia

Cataloguing-in-publication data

Mars, Ross

The Best of PAWA Volume 1: Selected Articles from the Newsletters of the Permaculture Association of Western Australia

ISBN 0 9587626 2 7 (set)

ISBN 0 9587626 3 5 (v. 1)

1. Permaculture 2. Organic gardening.

I. Mars, Ross. II. Willis, Robert, 1949 - III. Permaculture Association of Western Australia. IV. Title: PAWA Newsletter.

635.0484

Candlelight Trust

4110 Phillips Rd, Mundaring. WA. 6073.

Ph/fax (09) 9295 1933

The Formation of PAWA

Bill Mollison can be credited with starting permaculture in Western Australia, as is true in most places. Ray and I listened to him at the national Down To Earth ConFest near Canberra in January 1978. We were already keen gardeners and permaculture seemed to be an excellent idea.

While we were away, Bill Mollison came and spoke at the local Down To Earth ConFest at Cambray, near Nannup, and more West Australians heard and liked his ideas.

Most of the people who were to become founding PAWA members already belonged to the Down To Earth Association which organised the ConFests. The idea had germinated, and a group of people interested in forming an Association met at the Environment Centre, which was then in Wellington St, Perth.

The first people involved were Dennis and Jog Thiele, Ray and Ros Hart, Denis McCarthy and Cheryl Lange. Denis and Cheryl were organisers in the Down To Earth group and they ran the Moontime School, which was a large, alternative, adult education school, where many like-minded people met, either as teachers or students, or both.

This was also our first contact with David Noël of the WA Nut and Tree Growing Society (now WANATCA), Naomi Bannister, who became PAWA's first editor, and members of the Organic Growers Association. These were exciting times.

The gestation of PAWA was short. Ray was the founding Convenor. The first Newsletter came out in November 1978. By that time, the seedbank and library had been started and the Association could offer monthly meetings and field days. Bill Mollison had returned in September to encourage our progress with radio programs, publicity and talks organised by Down To Earth.

We often had experts giving talks as we were all on a steep learning curve. The practical permaculture nights were very popular, as were field days in backyards and orchards.

The Newsletter featured local articles because most published information related to either eastern Australia or to the northern hemisphere. Local information was scarce and we aimed to do something about it.

We found that local enthusiasts had a lot to offer. For example, Paul Sinclair had a consuming interest in poorly known permaculture trees such as yatays, macadamia, carobs and candlenut trees, and he led us on walks to see mature specimens of these trees fruiting in Perth.

The rest is history.

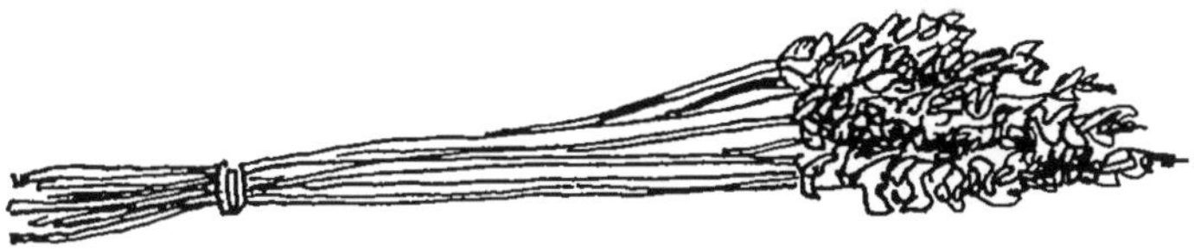

Forward

Bill Mollison

Director, Permaculture Institute

Permaculture is being embraced by people throughout the world. Some twenty years ago it was just a word, an idea about growing sustainable perennial crops. Today it is changing the way people live.

The Permaculture Association of Western Australia has been one of the fastest growing local permaculture groups in the world. Over the years, it has promoted the ideals of permaculture and relayed a message of hope to a vast number of West Australians, through its well-known and professional newsletter, and to many others throughout Australia and overseas.

This collection of articles, reprinted from some of the earlier editions of the PAWA Newsletter, represents a solid balance of theory and practice to help any reader understand the principles and concepts of Permaculture.

I am sure you will use this book time and time again, and it will become a useful addition to your reference library. Congratulations PAWA on nearly twenty years of service to the Permaculture community.

About Permaculture

What is permaculture?

The word *Permaculture* is a combination of the words *permanent* and *agriculture*. Bill Mollison, the co-founder of permaculture, says that it also means *permanent culture* as culture cannot survive without a good, sustainable agricultural base and a land use ethic.

The **Permaculture Association of Western Australia** (PAWA) exists to help people learn and use permaculture in their lives. The Association provides the opportunity for members to obtain information and build skills which are necessary to implement permaculture designs by:

- disseminating information and resources
- promoting the design and construction of Permaculture systems
- providing opportunities for interested people to meet
- seeking contact with any other related groups.

Other Publications by Candlelight Trust

All of the following are available from Candlelight Trust, 4110 Phillips Rd, Mundaring. Western Australia. 6073.
www.cfpermaculture.com.au

Books

1. Mars, Ross and Mars, Jenny. (1998). *Getting Started in Permaculture.* ISBN 0-646-20090-9. 100pp. (Revised from 1994 ed.).
2. Mars, Ross. (1996). *The Basics of Permaculture Design.* ISBN 0-9587626-1-9. 170pp.
3. Mars, Ross. (Supervising Editor). (1996). *Permaculture Volume Two: Best of PAWA - selected articles from the Newsletters of the Permaculture Association of Western Australia.* ISBN 0-9587626-8-6. 78pp. Now as e-book.

Videos/DVD's

1. *Passive Solar Design of Buildings.* ISBN 0-9587626-4-3. 23 mins. Narration: Peter Holland.
2. *RAPS - Remote Area Power Supply.* ISBN 0-9587626-6-X. 28 mins. Narration: Verity James.

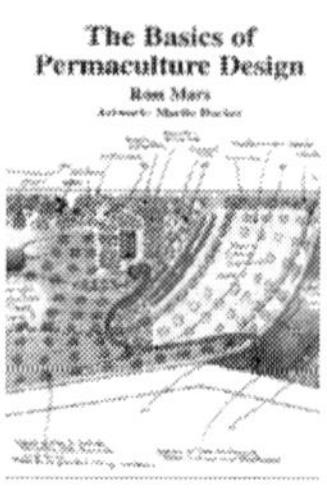

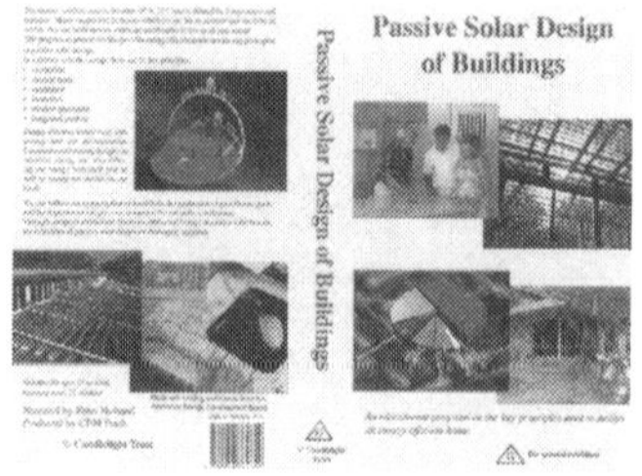

Contents

Permaculture - What is it?

The aim of permaculture is to link you with your food producing system. This can only be done by producing food yourself, thus taking responsibility for your own existence. Food and primary products can be grown in and around houses, backyards, street verges, parks, schools and vacant land. This applies to cities, towns and rural areas. These sorts of areas are part of a human ecosystem already, so why not fill them with useful plants as well as ornamentals?

The establishment of a permaculture system involves two inter-related steps: choosing the species and designing the system. But before starting with these, it is important to have sorted out exactly what you want from your particular permacultural system; that is, what food and other products suit your specific lifestyle and needs.

With this in mind, the first step is to choose the species of plants and animals that are suited to local conditions. These may come from anywhere in the world and should have as many functions as possible. A chicken in a permacultural sense acts as an egg producer, garbage disposer, tractor, meat source, pest destroyer, manure maker, glass-house heater, and much more.

By choosing a variety of species (and there are thousands to pick from) we can produce just about any primary product we want, such as food, medicines, shelter, fuels, oils and craft materials.

The other step is to design the system. Maximum use of space can be made with a complex planting of trees, understorey, shrubs, herbs and climbers. Plantings can be arranged to use every micro-climate available (such as walls, pond edges and hill slopes) and to create micro-climates for even greater diversity.

External factors such as wind, temperature and fire can be controlled positively by the arrangement of plants, roads, buildings, ponds and fences. In the same way, we can use plants to heat and cool buildings.

The strength of permaculture lies in the diversity of plants and animals used. Stability results from the inter-dependence of any one species upon others. For example, climbing plants may use the trunks of larger trees as support and at the same time provide shelter for predatory insects, and chooks benefit from foraging under fruit trees and also help to control fruit fly. This is in marked contrast to our current agricultural system where large areas are planted with one species, thus providing plenty of food for pests and little shelter for their predators. These monocultures can only be maintained by the use of artificial fertilisers and pesticides, and even then they may fail. The diversity of permaculture plantings means that there are always other crops to fall back on if one should fail.

As suggested above, low maintenance is another key factor of permaculture. This is partly the result of the use of perennial crops, such as nut trees that may produce for centuries and fruit trees for decades. It is also built into the system. For example, we can make plants and animals do the maintenance work for us. Chickens and pigs clean up dropped fruit and can plough the land, and plants can be self-seeding, produce green manures, shade other plants or filter water.

Permaculture plantings can be fitted to any scale in urban or rural situations. Actual designs will be as diverse as the needs and preferences of their designers and the local conditions. It takes time to establish, but once this is done the advantages increase - the system becomes more and more productive, there is an increasing range of products available, maintenance decreases and the system becomes more complicated. We may eventually be able to sit back and do less, admire our handiwork and think about something else.§

Applied Permaculture

Permaculture is an idea, or rather a set of ideas, which give rise to a technique of gardening and agriculture. The basic ideas of permaculture include:

- Diversity - within plantings, in the layout of different plantings, and in the total number of useful products.
- Complexity - there is a multitude of cross-linked functions which contribute to the overall stability of the system.
- Energy efficiency - productivity or efficiency should be at least partly based on energy, both internal, such as the percentage utilisation of sunlight, and external, such as energy subsidy in transport or processing.
- Minimal maintenance - the ideal of agriculture is that we do not have to work too hard or too long.
- Permanence - systems are designed to last with minimum input and high productivity, but with no gross net damage to the system. For example, systems cannot use up nutrients in soils for high yields and quick returns.
- High productivity - this is inherent in the complexity and diversity of the system if we design it properly.
- Human based designs - any value thought worthwhile can be included in the design, such as high diversity of local species, or wilderness preservation.
- Deliberate and rational planning - systems are actually designed within the framework. For example, a high local diversity is deliberate, even at some cost to the efficiency or productivity of particular items.
- The use of any modern techniques if they are appropriate - such as rapid breeding of new types of plants or animals to extend their range.

- The use of animals and plants to do things - such as poultry for making compost and plants for cooling.
- Self-perpetuating and maintaining systems - people should not have to be continually replenishing the system. For example, put in place self-feeding animal systems.
- The maximum diversity of products locally - this is partly for energy efficiency, but also for maximum local control and independence.
- The use of organic methods - these are the only sane methods, using natural predators for pest control, companion planting, compost and mulch.

The permacultural system of production can be made to work by simply doing it, and without having to battle anyone. It seems the easiest, simplest and most practical. For convenience, let us talk about permaculture as applied to six types of people, and engage in a bit of speculation about what it all means and about the priorities we need to establish in seeking a permacultural system.

1. **Self sufficiency**. Permaculture has more immediate benefit to people into self sufficiency than for any other group. The traditional aim of self sufficiency is to be largely independent of the system, usually with a concurrent idea of a simpler and more sensible lifestyle. The major problem is that this often approximates sheer hard work. The hard and healthy life can wear you down if it comes in excessive doses. The other problems are time and the necessity to continually watch the system, particularly animals.

Anyone can see the application of permaculture to these problems. For example, if you decided to have a holiday and it is mulberry time, well too bad - let the poultry have the mulberries. There will be something else in season when you return (apart from fantastic eggs and fat birds with purple faces). Perhaps the most attractive feature is the element of design - to build a self sufficiency permaculture will require skill, knowledge, imagination and inspired originality. These are the exact features that will be most attractive to the genuine self-sufficiency follower.

2. **Broad-acre Farmer**. This group will certainly not become extinct in a permacultural society. There are some products that are best grown in large quantities and do not need to be near the consumer. The most dramatic change would be the end of massive farming for a few cheap products for export. These will be replaced by products for more local consumption. The obvious examples are: plants for manufacture of fuels; plants as sources of chemical raw materials, for example, for plastics, solvents and oils; plants for fibres (timber, wood, paper); animals managed on extensive range with low maintenance, such as cattle on legume trees and pasture; plants for production of exotic chemicals, such as drugs and chemicals present in small quantities; and plants for production of craft materials.

These systems are unlikely to be as diverse as a permacultural garden. However, they will need to be run on a permanent soil-nutrient replacement basis. This can be done by taking only nutrients which can be replaced in situ (carbon, hydrogen, oxygen and nitrogen), by recycling from the consumer, or by fertilising, such as using sea-weed for minerals and trace elements.

Tree crops are particularly suited to this form of production because they are cheap in the long term (for example, no ploughing). In marginal land they are probably the only feasible and desirable cultivation apart from self-perpetuating pastures.

3. **Near-urban Farmer**. There will be a lot more of these. They will produce the enormous range of fresh produce which is the basis of a permacultural diet. Instead of producing a few crops in large quantities for storage or processing (for the rest of the year) they will produce a great range of products throughout the year. This reduces the peak seasons of activity and removes the necessity for storage or preservation. The diversity of plantings is used in reducing pest problems, so that, for example, fruit fly-prone fruit is broken up into blocks by resistant crops. The diversity of crops also gives a greater stability of output - in bad weather, for example, not all crops are affected the same way. The complexity of the planting increases the yield, and if these are trees, vines, annuals and animals in the same patch the total productivity is very high.

These permacultural systems are in the early years of development, and they can be considered as bigger, less diverse and less complex versions of the self-sufficiency systems. Although simpler, they will have to be carefully constructed for commercial production. The same considerations of fertilising apply as for broad-acre farming.

One possible system is fruit trees, bushes and vines in high diversity; perennial or tree legumes to fix nitrogen and provide a yield; annual crops to provide nitrogen, organic matter and yields; animals for pest control, such as poultry to clean up fallen fruit; composting and shredding, for example, chickens for light work and pig for major earthworks; and waste disposal for excess or damaged produce.

4. **The Urban Farmer**. This farmer will operate in urban areas tending crops which are probably on someone else's land, such as on a road verge, parks, vacant land and unused backyards. The simplest cases are fruit and nut trees as street trees, which can also be useful for shade, noise control and air quality. Legal and administrative technicalities present little difficulty in the long term. For example, leasing vacant land or unused backyards for vegetables is a simple step. Other problems include the pollution of urban areas, best typified by lead from petrol. Another idea is to grow non-edible products, such as bamboos for craft or structural use.

5. **The Micro-producer**. Here I include people (urban and rural) who are not primarily concerned with commercial production, but want to grow something either as a hobby, to save money, or because they want a particular product, such as flowers, very perishable fruit, rare things, craft materials, or unusual honey flavours. In many of these endeavours the cost or effort involved is no problem. A permacultural system will do several things for these people, such as greatly increasing the range of useful plants and animals available, and providing useful ideas like poultry as compost makers.

6. **The Consumer**. This group doesn't produce primary products because they may not like gardening, are itinerant or for some other reason. Permaculture has a lot to offer the consumer, which includes all of us, of course. Food will be fresher, a lot more diverse, better quality and produced locally. If it has been processed it will have been done locally and recently, so it will not have to be loaded with preservatives. It will be a lot healthier as a result.

Instead of going to the supermarket to pick from hundreds of highly processed items, and picking up a few fruit and vegetable items, the consumer will be faced with a stunning variety of fresh fruit, fresh vegetables, animal products, nuts, grains, beans, flavourings, and processed goods. There will be considerable seasonality in diet which should be interesting. No one will miss what was in season last month because other things will have arrived.

When it comes to designing a permacultural system, one is faced with a very formidable task. The first priority is to decide just what you are trying to do. The second step is to integrate your particular tastes. If you are building a self-sufficiency garden and you don't like olives, leave them out, even though they thrive here. If you have an ambition to be a commercial fruit grower, think about diversity and fruit trees primarily.

On the practical side there are many problems which are of immediate and urgent interest to all permaculturalists. Major priorities include:

1. The diversity of useful plants and animals available locally needs to be increased. We need trial plantings of anything and everything. We also need to ask for official action, particularly in breeding programs for local conditions. However, one thing we don't need is a greater diversity of pests and diseases, so careful consideration needs to be made of this.

2. We need considerable discussion, thought and research on the design of real permacultural systems. Groups of people are often best because of the diversity of ideas and knowledge.

3. We need access to experts and more experts. The easiest way to find or make an expert is to ask questions. In particular, there are already extensive government services available. We should be using these, and encouraging them.

4. We need a lot of work on organic growing, particularly for commercial use. We all want to use organic methods, but in most cases they are poorly understood, very badly documented, and untested or impractical for commercial use.

5. We need to spread information and resources so that they are available to anyone who wants them, and to encourage people to get involved in decision making.

Hopefully, the more people involved the better the answer, particularly if they are well informed.§

Considerations When Designing Small Animals into a system

In looking at the animals most suitable to an urban situation (zones one and two in a permaculture design) we are dealing with small livestock which may be applicable also on large acreages. When considering an element, we list three major factors. These are Needs, Products and Functions. We will use the chicken as an example. The chicken is the most widely used animal in permaculture systems and is as applicable to 1/4 acre (1000 m^2) as it is to 4000 acres (1600 ha).

There are perhaps two other considerations; Breed Intrinsics and Legalities. There are often laws prohibiting the urban enthusiast from keeping various animals. Whereas there are many restrictions on poultry, bantams are usually classed as a cage bird and it is often possible to fit them in where no other chicken is allowed.

The perfect system maximises all products and functions and provides the element (in this example the chicken) with all its needs with a minimum of work from you. Obviously this is a goal that is rarely obtained, e.g. if you are using the chicken to turn compost in a small compost area, you may have to supplement feed. Chickens should have a larger range of forage plants, such as tagasaste (seed and leaf), coprosma (berry on female plants and leaf), fat hen (seed and leaf), mulberries, cotoneaster (berries) and acacias (seed). Many of these plants make good shelter belts and nurse species in an orchard situation.

Dust baths must be kept dry and can include ash, lime, diatomaceous earth, powdered charcoal and dried/powdered aromatic herbs such as wormwood, pennyroyal, tansy and pyrethrum. Most of the functions and products of the chook remain unused in a commercial poultry battery and consequently become pollutants. The permaculture definition of pollution is an unused product. Even manure has to be removed. It's much better if you can have the chook scratch the soil, manure it, de-weed it and prepare it as a garden bed. We do this either using small mobile tractors or by establishing a series of pens around a central roost, where each one rotates from chicken pen to vegetable garden.

If you plant deciduous fruit trees in these pens, you will further increase the product of your garden/chicken system and provide some summer shade areas for appropriate vegetables. Heavy rocks around the feeder root zone will stop chooks from disturbing these. If you can't access rocks, then boards or even chicken wire stretched out and staked with tent pegs will work. When this area is garden, you can grow all kinds of annuals between the rocks - such as lettuce, broad beans and peas. Vines on the fences will further increase your yield from these areas and provide shelter from wind and sun.

A permaculture principle is to make every element serve many functions. So, if you site the roost in the corner of a hothouse, the body heat of the chickens, as they roost at night, will increase the temperature inside the hothouse and also contribute CO_2 to the air. One of the big problems of hothouse culture is replacing CO_2 which the plants have converted to plant food and oxygen. Some commercial hothouses actually have to introduce bottled CO_2.

Another Permaculture principle is to serve every function with many elements, so we may decide to add a water tank to the hothouse as well. The water in the tank acts as thermal mass, storing the heat of the sun and releasing it at night. We can stock the tank with appropriate fish and water plants. The water from the tank is high in nutrients for the garden and so promotes growth in plants.

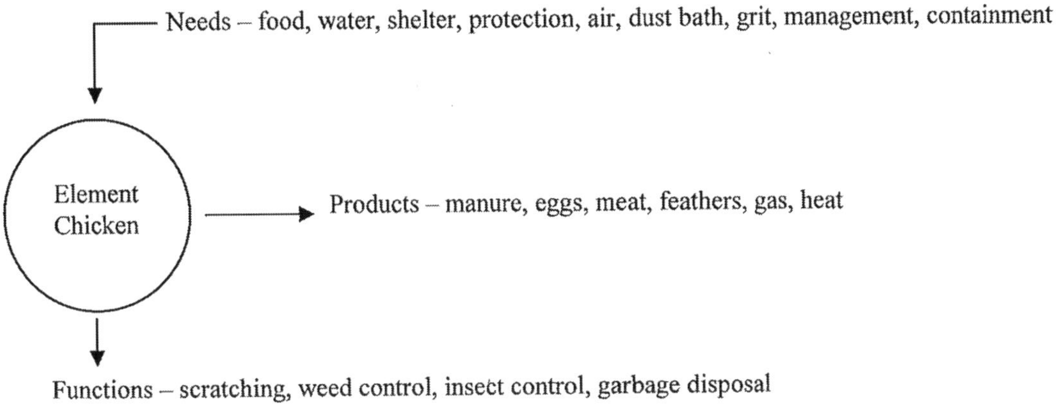

Water plants such as azolla and duckweed are high protein feeds for ducks, geese, chooks and pigs. Azolla, a type of water fern, has a bacterial associate which lives in the azolla leaves and fixes atmospheric nitrogen. The nitrogen does not become available until the azolla dies and drops to the bottom, breaks down and becomes available to the next generation. If the azolla, which reproduces very quickly, is harvested, it can be applied directly to the garden as a mulch which breaks down to a peatmoss-like soil.

By starting with any element, we can build up an extremely complex integrated system. By listing the needs, functions and products, and understanding the breed intrinsics of each element, we can fit whole systems together which are quite sustainable.

An early Agriculture Dept. bulletin stated that all successful orchardists kept between 30-50 chickens to the acre for complete control of weeds, fruit fly, codling moth and for total manure supply. When the chook was moved into the battery, the orchardist lost these functions and was forced to use chemicals as a substitute. Many of these orchards are now so contaminated that poultry can't be run for decades to come.

A complication occurs when poultry range on such contaminated soil. In the course of scratching in the soil and picking up soil particles, the birds ingest some of the residual organochloride, or other poison, which persists in the soil for many years. If you don't know the pesticide history of your soil then have your eggs tested. You may be staggered by what you've been eating and assuming to be good organic food.

As with hybrid seeds, the next generation does not breed true to its parents, and consequently each generation has to be bred from those specific breeds. Before the battery hen development we had an extremely diverse gene pool of poultry, with flocks being kept on farm and in backyard situations. Those birds were constantly being selectively bred for utility. For example, a breeder of meat birds will always cull the smallest birds.

Today, the few remaining backyard flocks are usually the cheap and readily available first cross, usually secured as "burned out" battery birds. The people who have kept the breeds cross and select birds for feather, how the bird looks, whether the comb is as it should be, correct markings, colour of legs and so on, and pay little or no attention to utility qualities. We have lost in the last half a century what it has taken thousands of years of selective breeding to arrive at.

Even the major commercial breeds such as White Leghorn, Australorp and New Hampshire - the components of the popular first cross - do not have a very broad genetic base. Too often, the fancy breeding stock has been used. We should be trying to piece together some of the rarer breeds that have good utility and quality and selectively breed for these qualities. Initially, this could mean a lower production, but we have got to start now before it is too late.§

Permaculture Gardens – Another Kind of Garden Living

Garden living is an integral part of our species' cultural heritage. Our ancestral home was a natural garden. Since the earliest times of recorded human settlement, people have made and lived in gardens of one sort or another.

Gardens remain an important cultural link with our past. They may well be a practical necessity for survival for the present and the future. Permaculture gardens are quite different from any others and no two are alike. Each one reflects the lifestyle and preferences of their creators.

However, they all are complete gardens, needing very little from the outside world to sustain them. The products of these complete gardens are many and never-ending. Food, medicine, fuel and construction materials are there for the taking, in abundance. Permaculture gardens can also cater for the visual and spiritual needs of their creators.

What is permaculture? Is there a secret?

There is no magic or secret ingredient needed to create a permaculture garden. The master gardener, Nature, shows us the way. The science of ecology, which is defined as the study of the relationship between an organism and its environment, is used to study the workings of the natural world in a holistic way.

Permaculture gardening is based both on spatial (space) and ecological principles. It is a safe and highly productive perennial system, consciously designed by people for their own immediate and enduring benefit. Permaculture gardens are growing and can provide an abundant future for all mankind.§

Chokos

Choko vines are very productive. Chokos keep for ages in the fridge, so you can even have a winter supply.

To germinate, place the choko, stalk end down, on top of some compost at the site where you want to grow it. In winter to early spring, the oval-shaped lips of the seed will poke out of the end, and from between the lips a trailer will begin to shoot. This needs support, and I've used thin wire or bamboo sticks.

A site with part shade is good on the corner of a shed or house, or at the bottom of a tall tree! (a deep-rooted one that won't steal the choko's food), or near any kind of support you can put up - the iron grid base which is used in building is good.

CHOKO

While the fruit is waiting to shoot (getting pale and yellowish) it has plenty of food to start it off on its journey, and then it can feed on the compost or lucerne hay.

In the compost, use some blood and bone, bone dust and lots of lemon, orange, and grapefruit skins (for potash), and banana skins. About 10 g of Epsom Salts watered over it will also help. This mix should result in good NPK and trace elements. As it grows it can have boosters of liquid poultry manure, wood ash, and lots of comfrey "juice".

In its first year it may refuse to produce flowers. If you can cause the plant some mild stress it may trigger the flowering and fruiting cycle. For example, cease watering or chop some roots with a sharp spade, or cincture the main stems back - basically deprive it or attack it in some way to force it to provide another generation of fruits. By the early autumn after planting in spring, it should produce flowers and fruit.

In its second year it should produce a smallish crop in spring through to summer. The vine should last for about five to seven years, but after three years tends to get tired and become less productive - so keeping some large fruit for seed is important.

Chokos like to keep their roots cool and don't like heatwaves unless it's humid as well as hot. Lots of water is needed when it's hot, and they hate wind, especially when young, although the mature plant recovers quickly. Train your choko vine over fences and walls, and by doing this it will create shade as well as abundant fruit.§

The Humble Spud

Potato growing is a simple process. If you have a space in your garden in an open situation you should try it. Firstly, we will deal with the ground. Last year's chicken run is an ideal spot. Move your chickens to another run about two months in advance and then dig the run over with a fork. You are advised to do this on any patch you choose.

Planting times can vary from the autumn until the end of winter, thus limiting you to the cooler months when you need low watering levels. Plantings done in early winter will mature with natural rainfall. Plantings in late winter will give fresh potatoes for summer with watering to the end of the growing period, while autumn plantings will have to be watered from the word go until the winter rains come. Try a few during these periods to see which suits you.

When planting potatoes by conventional methods, rows must be dug approximately 125 mm deep. The width of row can be varied according to space, and somewhere between 50 and 100 cm, preferably 75 cm. Put a handful of blood and bone per metre along the bottom of the row, preferably on top of a little compost, then plant the potatoes on top of this about 35 cm apart. Fill in the rows, trample down and rake level.

Your potatoes will show through varying from 10 to 21 days according to moisture, temperature and stage of sprouting when planted. When growth is 100 to 150 mm high take one handful of blood and bone to use as a side dressing per metre of row. Ridge a little to cover the blood and bone. You can complete the ridging about two weeks later. This is done to prevent potatoes from growing through or too close to the surface and turning green. If you are a complete novice you will find it takes a little trial and error like most projects in the garden. Following this method you will get your seed 100 mm deep with a little compost in a 125 mm row. Be sure to ridge at the appropriate time and you will get no hassles.

You can expect nice new potatoes from ten weeks onward. This will vary according to the previous soil conditions. The crop should reach maturity in about sixteen weeks when the tops will start to yellow and die off. You may then lift the crop and store in a cool place. Stored in this manner they will keep in good condition for up to three months.

Some do's and don'ts

Do

1. Save small, damaged and green potatoes from your eaters - these are free "seed".
2. Your seed will vary from 25 to 50 mm in size. The smaller ones can be planted 250 mm apart because they make less tubers than larger ones.
3. If you have very large seed cut them up leaving two eyes on each piece. Allow to callous over before planting.

Don't

1. Plant in waterlogged soil where the area might flood, as a potato will rot if in water over 24 hours.
2. Put fertiliser directly above the potato seed, or it will burn off the chits as they grow.
3. Let snails over-run your plot, as you will get no "bottoms" if they eat the tops.

Good luck to all prospective growers.§

Edible Weeds

Edible weeds? You might ask - isn't that a contradiction in terms? However, a weed is really just a plant out of place and as weeds are so successful and easy to grow, surely we can find a place for them in a permaculture system. Many can be used to provide a useful supply of 'greens' especially from those odd, hard-to-cultivate corners of the garden.

There is much still to be learnt about eating 'weeds', but here are some of those we've learnt to incorporate in our cooking.

Purslane

Purslane (*Portulaca oleracea*) is our favourite summer weed, which grows well from seed. You will find it everywhere and it is even sold as a herb in the USA. This plant grows over summer from about October to May, dies in winter and grows up from seed the next year. Once you have one plant you are sure to have more the next year. It has tear-drop shaped fleshy leaves, up to about 1 cm long, reddish stems (its most distinctive feature) and spreads over the ground like a flat mat, rarely growing more than 5 cm tall. It has small, yellow flowers in February to April and produces tiny black seeds from March onwards. It seems to thrive on light sandy soil and loves a sunny position. It is one of the more resistant summer weeds, even growing in back laneways.

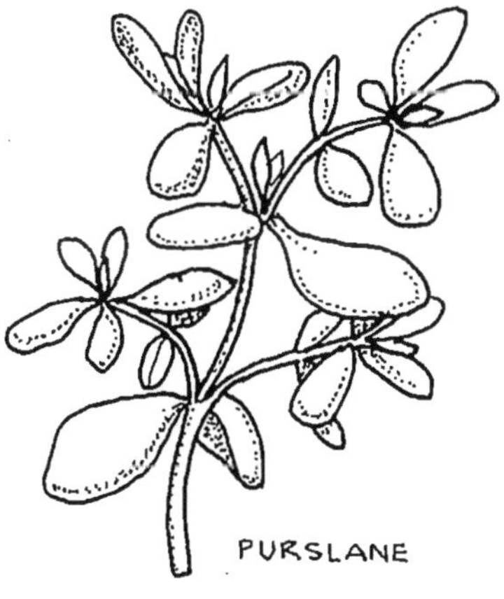

The young leaves and stems are really nice - chopped and used raw in summer salads - they have a slightly nutty, cucumbery flavour and the tiny black seeds are a nice addition when they appear. Continual cutting of the tips encourages more tender young growth.

Chickweed

In winter our favourite (and also the most prolific in our garden) is chickweed (*Stellaria media*). The name chickweed seems to be used for other things as well, so be sure you are on to the right one before eating any. Chickweed starts to grow from seed in May-June once the ground is wet and cool from rainfall. It has a very fragile root system and trails over the ground with pale green, delicate, almost hairy leaves and delicate tiny, white flowers. The stem tips finish in distinctive four-leaved heads which you can quickly learn to recognise.

Chickweed grows 15 to 25 cm tall, possibly more if there is something it can ramble over. It prefers to grow in shady, protected positions, so we encourage it to grow under our fruit and citrus trees as it needs no cultivation. It is much easier and faster to harvest if you keep the area free from other weeds, so you can cut handfuls with scissors and not have to sort through for 'foreigners'.

Chickweed, on its own, can be lightly steamed and eaten like spinach. We find it an excellent addition to omelettes, stir fried vegetables and rice dishes. In fact, it is a good green addition to any vegetable dish and is quite tasty raw as well as cooked.

Stinging Nettle

Another good winter weed to eat (although I wouldn't go out and get this for my garden unless it is already established there) is the common stinging nettle (*Urtica spp*).

The tips can be carefully picked by hand - it can be done without being stung, although it may be safer to use gloves and scissors.

Cooking destroys the 'sting' - this is not a weed I recommend raw! It can be steamed lightly - the taste is very much like spinach again.

Picking the tips will cause the plant to put out lots more tender new tips for eating as the old leaves are really quite tough and bitter, and not good to use.§

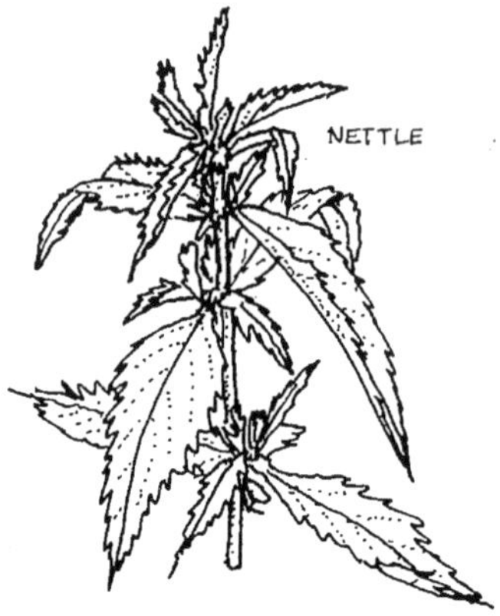

Herbs

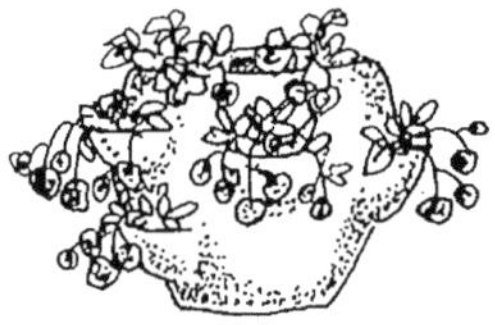

Introduction

One of the earliest known records of the use of herbs was documented by the Sumerians in about 3000 B.C. and lists the use of laurel, caraway and thyme. As well as having extensive uses in cooking and religious rites, such as embalming, the history of herbs is one linked with mystic, magic and superstition, and this latter aspect of the history is undoubtedly related to the use of herbs in medicine. Prehistoric humans used plants to treat physical complaints - the process of discovery would have been by trial and error. Nowadays, the documentation about herbs is extensive, but research and writing still draws reference from the early writings of such people as William Turner, John Gerard and Nicholas Culpepper from the sixteenth century.

If we are looking for a definition of a herb, it is possible to classify them among the enormous division of plants producing oxygen and transforming the simple elements and compounds of the soil and air into the complex substances animals and humans need for food. The plants of primary interest to us are classified as angiosperms and gymnosperms. Angiosperms are the flowering plants, a group numbering over 20,000 species, and gymnosperms are the trees and shrubs numbering over 500 species. Herb was the word once used to describe any growing plant and the distinction between herbs and vegetables was not widely used until the 17th century. The most likely usage of the title 'herb' nowadays would apply to a plant which is used for flavouring in cooking, also for its aroma and medicinal properties.

To place herbs in the context of Permaculture is not difficult as they, by definition, have immediate and obvious uses in the garden for companion planting, as an insect deterrent and as a food source. The ideal growing place for the majority of herbs is close to the house for easy access. From experience, I have found that the more crowded the garden with herbs and vegetables, the better things progress and the greater the deterrent to bugs. The traditional herb garden plan is one given over exclusively to herbs, but to mix the vegetables in with the herbs appears to be a successful alternative. Obviously, you must know the root system of each plant in order that they are not competing for the same nutrients, but, in general, the more the better and the wider the variety works well.

Diversity is, of course, important in a permacultural setting and herbs can provide good return for the minimum of upkeep. My first herb garden was started on a small sandy patch of soil which was so poor that even after mulch and compost and lots of seaweed had few worms, but this did not stop me from successfully growing about 30 different varieties of herbs.

There are three major botanical families into which herbs are grouped:

- Labiatae - the mints, including sage, rosemary, thyme and marjoram with whorl flowers and simple leaves.
- Umbelliferae - the parsley group with long cylindrical stems and umbrella-like flower tops.
- Compositae - the daisy group with ray or disc flowers.

The first and second parts of the botanical name identify the genus and species of the plant, the species being the smallest group that can be consistently identified as distinct from other groups. The genus includes various species that show certain common features. The two names are usually derived historically, from some physical characteristic, use of the plant, geographical location or from some botanist being honoured. The species name can give an indication of the most historical, important use of the plant, e.g. officinale tells us that the plant has been used at some time as a medicine; wort also indicates medical use.

Propagation

Annuals generally flower, seed and die in one year, as with basil (*Ocimum basillicum*). Biennials grow over 2 years, going dormant after the first year, flowering and seeding in the second year. Perennials grow 3 or more years and much longer in many cases. Apart from one or two herbs such as chervil (*Anthricus cerefolium*) which cannot tolerate being moved, most herbs can be sown in seed beds, and in a warm climate this means any time of year, although in colder areas protection under glass would be needed to promote germination.

Care

It is important that herbs have good drainage. This does not present much of a problem in sandy soil but in heavier type soils it is important to keep this in mind. Plenty of mulch will, of course, cut down the need for watering, although herbs such as comfrey (*Symphytum officinale*) require constant watering. Wind damage can be quite a problem and quite a few of my taller herbs show signs of leaf burn.

Harvesting and drying

It is essential to follow a few important steps in order to retain the desirable qualities present in the herb. These can be outlined as follows:

1. Always gather herbs in the early morning after the dew has dried and before the sun has dispersed their volatile oils.
2. Do not pick when still damp after rain.
3. Cut with a sharp knife except for chives, which should be cut with scissors.
4. Handle as little as possible to avoid bruising.
5. Dry in the shade to conserve the oils as much as possible.
6. Keep separate from other herbs to avoid cross-flavouring.

The best drying methods are:

a. Spread in a thin layer on an old fly-screen for the best air-flow.
b. Hang in bundles for a few days, then wrap in paper and store in a dark cupboard, or better still, an airtight container.

The herbs should be brittle and dry before they are stored. Light, moisture and air are the things which will destroy all your hard work. The practice of keeping dried herbs in clear glass containers in the kitchen is about the worst way they can be stored. The containers should be dark, and if this is not possible, then stored in a dark place.

The best time for harvesting is generally just before flowering when the volatile oils are at their peak.

- Leaves - pick as above, discarding any damaged ones. Keep leaves attached to the stem. When dry, remove from the stem and discard the stem as this is not needed. Only ever take about one-fifth of the leaves if the plant is to be retained.
- Flowers - pick when fully open, select only those that are unblemished.
- Berries - gather when fully ripe and glossy.
- Seeds - cut stems with seeds on as soon as they have gone brown. A paper bag over the head of the seed pod will prevent loss.
- Roots - gather in the autumn.

Herbs will generally retain their potency for only a year, except for bark which will be viable for at least three years. The more finely powdered they are, the quicker they will lose their taste.

Freezing

Herbs have a freezer life of about six months. This method is particularly useful for the soft-leaved herbs such as mint, chives, parsley, balm, fennel, basil, dill and sorrel. After harvesting, wash and shake dry. For short periods of storage, wrap immediately in plastic and freeze; for longer periods, blanching is advisable. Place in boiling water and then in chilled water. They can then be wrapped and frozen or chopped and packed in ice cube trays with water; remove when frozen and store in bags.

Herbal oils

Herbal oils are good for cooking and cosmetic purposes. The leaves should be picked from those herbs in flower and sunlight is the necessary ingredient for processing. Crush the freshly cut herbs with a mortar and pestle or put in a grinder. For 300 mL of oil you need two tablespoons of the pounded herb at the start and the same amount again in three weeks.

Fill the clear glass container three-quarters full and add one tablespoon of wine vinegar and three or four black pepper corns. Place in full sun all day and shake twice daily.

After three weeks strain off the oil, squeeze oil from herbs and discard the solids. Add two tablespoons more of herbs to the oil and repeat the process for another three weeks. It is possible to make herbal oil when there is not enough sun by gently heating oil in the bottle each day for a few hours. At the end of the six weeks exposure or heating, strain the herbs again, bottle the oil and add a sprig of the herb used for decoration and identification, label and date. These oils can be used in cooking and salads. If you wish to use the oil for cosmetic purposes it can be used on the skin or rubbed into the hair the day before washing. This softens the hair and reduces the drying inevitable with hard water and a dry climate.

Herb vinegar

This is made from the leaves. Use two cups of fresh leaves to five cups of vinegar, add two or three peppercorns and shake every day. After ten days, taste; if not flavoured enough strain out leaves and renew, leave for another ten days. Strain and bottle with a fresh herb sprig.

Seed vinegar - two tablespoons of seeds to five cups of vinegar. Bruise seeds, pour warmed vinegar over them and put in warm place for two weeks, shaking occasionally. Strain and bottle.

Garlic vinegar - put garlic cloves into vinegar, leave 24 hours and remove garlic.

White vinegar goes best with salad burnet, basil and tarragon; cider vinegar with mint and red wine with garlic. It is possible to make pot-pourri, pomanders, herb pillows, sachets, and lavender and tansy bags to perfume clothes and deter bugs respectively.

Herbs for health

The validity of the knowledge gathered by the ancient herbalists has been amply verified by the commercial use of such herbs as foxglove, from which digitalis is extracted for use in the treatment of heart conditions. There are many other examples of this kind. Self-medication should not be undertaken without considerable knowledge of the effects of the use of herbs. The effectiveness of herbs is a gradual action based on restoring the natural balance of bodily functions that constitutes health. There is certainly no such thing as an instant cure-all with herbs. There are several herbs that if used unwisely, for example, tansy tea, can cause severe poisoning. So be sure to check your facts and know that what you are using is safe.

Here are some of the ways to prepare herbs for medicinal and cosmetic uses with fresh herbs.

- Infusion - to make teas, place one or two teaspoons of the herb in a porcelain pot. Pour hot water over. Some herb teas, such as chamomile, require short infusion, about four minutes, whereas others may require longer.
- Decoction - used for roots, wood, bark and seeds to extract mineral salts and the bitter principles of plants. Boil approximately 15 g of herb to 1 cup of water for three to four minutes and leave to steep for two to three minutes. Hard materials need boiling longer (10 minutes); strain and use like a tea.
- Compress - soak cloth in infusion or decoction, wring out and apply to infected area.
- Poultice - bruise and crush herb to a pulpy mass and heat, wash wound with herb tea, such as chamomile, and spread cooled pulp on wound. Wrap with poultice on wound - for both humans and animals. You can make your own ointment or salve by mixing one part of the herb in powder form (about 125 g) with warm petroleum jelly; heat slowly, add 30 g of beeswax, a little benzoin gum or a drop of tincture of benzoin per 30 g of fat, which will help to preserve the ointment.§

Lemongrass

Lemon grass is a little known perennial herb from South East Asia, where it has been used for centuries as a culinary herb, imparting a delicate lemon flavouring to curries and other dishes.

The leaves are long, up to one metre, and fairly thin, light green in colour with a very characteristic lemony odour. The plant increases in size and diameter by stem multiplication. The leaves grow from stems about 150 mm above the ground and grow only over summer.

The plant lies dormant over winter, so the plant should be harvested at the end of summer. This involves cutting off all the leaves back to the 150 mm stems, but not cutting off the stems. The leaves are then dried and cut into convenient lengths for use.

The plant may be divided by separating the stems into pieces with roots. One stem planted in spring may yield as many as 20-30 stems by the following winter as well as several jars of dried leaves, so it is a good producing plant.

With some summer watering, lemon grass grows very well in sandy soil. Sunlight and water are the two most important factors influencing growth. A good thick mulch around the plant obviates the need for daily watering, although occasional watering in a dry summer is necessary. It does usually die back in winter.

Lemon grass is used as a hot herbal tea, perhaps sweetened with a little honey. In summer, fresh leaves can be used directly by cutting them into short pieces, covering with water and brewing 10 - 15 minutes. In winter, dried leaves are used - these are available commercially but are very expensive, in the light of how easily the herb grows. Note that dried lemon grass should be stored in dark, airtight containers, otherwise it loses its odour rapidly.

Lemon grass leaves can also be used in curries but, like bay leaves, should be removed before eating as they are horribly tough. Asian curries use the lower part of the stem, rather than the leaves. In summer, one of the most refreshing, and thirst-quenching, drinks I know is cold, honeyed lemon grass tea.

Lemon grass also combines well with other herbs or ordinary tea. Lemon verbena and lemon grass is an especially good combination. Lemon grass tea is good for clear skin and bright eyes as it contains large amounts of vitamin A.

Thus lemon grass is a very valuable permaculture herb - it is perennial, grows prolifically over summer, producing enough for summer use, and is easily dried and stored for winter use.

It makes a refreshing, cold summer drink, a pleasant hot winter drink alone or combined, and is a useful lemon flavouring in stews, curries and other cooked dishes.§

The Mulberry Tree

There is a strong case for including the common, backyard mulberry tree in a planned Permaculture system. The mulberry tree has been used for thousands of years by humankind, and is known for many uses.

Varieties and species of the Morus family are to be found in sub-tropical, temperate and cool climates, extending through Asia Minor to Europe, and in North America. In Australia, the mulberry is a direct descendant of the English garden tradition, and thus it is the common, black or English mulberry (*Morus nigra*) that may be found.

The black mulberry is a short, broad-leafed tree, growing six to ten metres high. It has a weeping spread of branches above a short rough trunk. Unisexual flowers are followed by oblong, short stalked fruit (actually a fruit cluster), about 25 mm long. Colour is from dark red to black. The fruit are very juicy, with a very sweet taste.

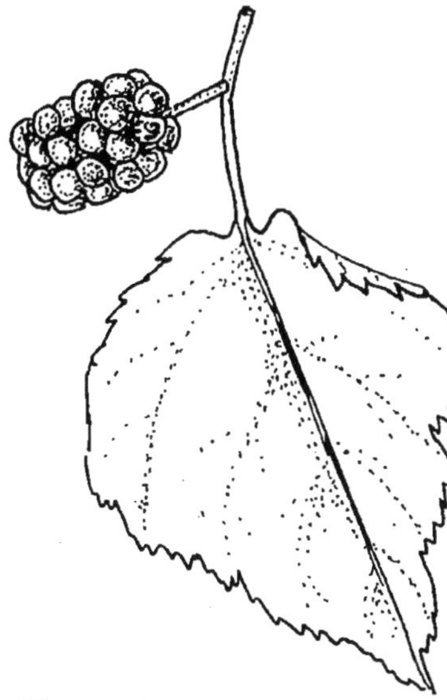

BLACK MULBERRY

The mulberry tree does best in warm, well drained, loamy soil, producing best crops when fed regularly with manure. It grows in sub-tropical, temperate and cool climates, so long as there are no late frosts.

The tree may be propagated most easily and cheaply by cuttings, up to 2.5 metres long planted 30 cm deep in well mulched soil. This should be done in autumn, when the cuttings will easily develop roots - faster if placed under glass. Alternatively, graft buds onto white mulberry wood. Young mulberry trees grow rapidly and are well able to withstand transplanting.

Plants bear fruit in their second year after cutting, increasing their crop yearly, and maturing after 15 years. Mulberries have a relatively long fruiting season, being from five to fifteen weeks each year. They are hardy and prolific growers, and should be cut back heavily in autumn to retain their shape. The sheer weight of its crop might otherwise cause the long weeping branches to snap.

The fruit are highly nutritious, containing glucose, protein, pectin, tartaric and malic acid, and a large reserve of grape sugar, or invert sugar, which may later be converted to starch. Eaten fresh, they are a mild laxative and are a suggested source of relief for sore throats when made into a syrup. Commonly, the black mulberry is a kitchen crop, being used in jams, pies and winemaking.

Since the fruit bear over such a long season, any commercial harvesting is too labour intensive, and the tree lends itself more readily to providing a no-labour forage crop for poultry or pigs. Picking is easy enough when considering domestic consumption, and a good tree will yield a bucket a day for many weeks.

The mulberry is frost resistant, being able to recover and bear a second crop if the primary buds are short lived. It suffers little from any pest, and has several companion planting applications. The tree is a good support for grapevines, although harder to pick. The extra height gives the vine good air circulation, resulting in less pest and mildew problems. Mulberries act as a good decoy crop for cherries and strawberries, drawing birds away, or alternately attracting them into the garden.

Through pruning, large quantities of firewood are produced. The wood is light, and burns at a low temperature. Fruitfly do not seem to be a problem.

There are two varieties of black mulberries. "Black English", which likes cooler districts, and "Hicks" which is more suited to warmer areas. They have different fruiting times.

The white mulberry (*Morus alba*) is also worthy of consideration. Originally a native of China, its use around the world has led to the development of many varieties, particularly in Japan, chiefly for the silkworm industry.

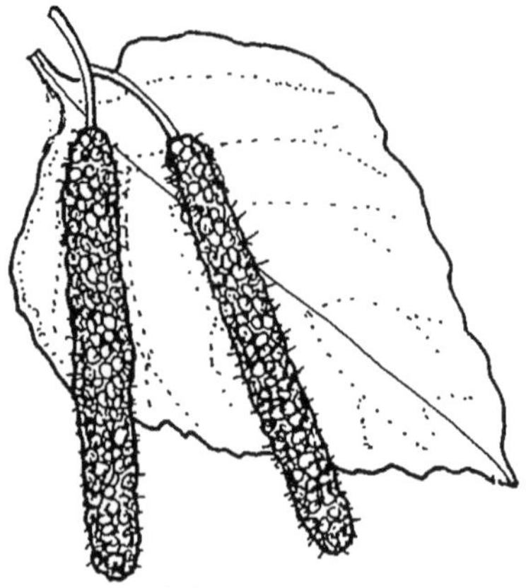

M. alba is a deciduous tree attaining a height of ten to fifteen metres at maturity. It is the favourite food of the silkworm, particularly due to a latex-like gum exudate, which is produced in the branches. This is thought to give tenacity to the fibre produced by the worm.

The American, or red mulberry (*Morus rubra*), is native to the American continent. Its fruit is less acidic than the black mulberry. It has been used in pre-industrialised agriculture as a labour free source of pig or poultry feed, utilising free-range principles.

Mulberries are an asset to any garden and can be pruned and managed, even in a small backyard. They are truly a very useful permaculture plant.§

The Carob Tree and its Products

The carob (*Ceratonia siliqua* - family Leguminosae) is an evergreen tree with rich glossy foliage, which is indigenous to the Eastern Mediterranean regions. It is best known for producing long, brown, bean-like pods which form a valuable addition to the diet of both farm animals and people. Carob trees have had a very long association with people, at least 4,000 years with Eastern Mediterranean peoples. Remains of old carob and olive groves planted by the early Romans have been found in Tunis.

Carob trees are especially valuable in any Mediterranean permaculture system, because as well as providing nutritious food, they are excellent shelter trees. They grow well in a wide variety of soils, from beach sands to heavy clay loams, but must have reasonably good drainage and soil aeration.

Young trees are often frost susceptible but seem to vary in their individual resistance, while adult trees are much hardier, recovering from frost because of their large root system. The tree relies on a very deep tap root (20 m has been recorded) but it also has an extensive lateral root system lying near the surface to utilise gentle showers which do not penetrate deeply into the soil.

Trees usually reach a height of 10-15 metres and resist drought and hot winds. However, although drought resistant, the tree will barely subsist on an annual rainfall of 120-200 mm and then only for a few years. On 250-300 mm in deep soils it will grow quite well and even produce light crops of thin pods. Only where winter rains total 500 mm or more, on the average, should commercial carob culture be attempted without irrigation.

Carob trees come in three types: male, female and self-fertile (male and female on the same flower). Self-fertile trees tend to be poor producers. The scent of the male trees in bloom is sometimes offensive and usually very strong while female trees have very little or no odour.

Pollination is effected by bees, flies, small wasps and night flying moths. The effectiveness of wind-borne pollen is not known, but isolated female trees have been known to produce light crops.

One male tree to 8-10 females is adequate, but it is important to choose a male tree with a long blooming period to provide pollen for both early and late flowering females. Pods take a whole year to mature, so last year's pods are ready for harvesting at the time when this year's flowers are in bloom.

Some trees drop their pods when mature, while others hold their pods tightly and these have to picked. At the end of the autumn it is important to remove all mature pods from the trees to prevent pests overwintering in pods and infecting the next year's harvest.

Whole pods are excellent feed for even heavy draft animals or as a fattening food for steers and pigs. For milk cows and goats, it is necessary to feed a protein supplement and additional roughage. All animals will eat the pods greedily once they have acquired a taste for them. A few pods can be broken into their regular feed to introduce the taste. For that matter, many human mothers have teethed their babies on carob pods which gives the children an early taste for carob.

Carob as human food is available in many forms. Raw pods are delicious as a snack. Carob powder, which is very fine, can be used as a cocoa or chocolate powder substitute in cakes, drinks, custard and biscuits.

Solid carob "chocolate" is a good energy food. For example, on bushwalks you can roast then crush the pods, percolate or boil these and you have a delicious hot carob drink.

Recipes for carob cooking are becoming more common, and you can substitute equal volumes of carob powder for the chocolate in your favourite recipes. Incidentally, carob powder has long been used as a cure for diarrhoea. Two heaped tablespoons of carob powder in a glass of hot water is very effective at stopping severe cases of diarrhoea and yet eating carob in food does not cause constipation.

Carob seeds are the source of monogalactan gum, which is also known as carob gum. This is used for paper sizing, photographic emulsions and stiffener for ice cream, salad dressing and pie filling. There are literally hundreds of industrial uses for carob tree products. So if ever a tree had diverse functions it's the carob. It provides windbreaks, shelter and leaf litter to enrich the soil, produces food for stock and people and is a beautiful tree into the bargain.

Carobs are usually grown from seed. If the seeds are removed from the pods as soon as they are mature, but have not dried out, they will germinate readily. This stage is determined by the stem of the brown pod still showing a bit of green colour at the end.

Once the pods dry thoroughly, the seeds become flinty and special measures must be taken to make them germinate.

The seed may be scarified by clipping or rubbing with sandpaper on the broad end, soaked for 24 hours then planted. Otherwise hot water treatment is easier and works equally well.

This method involves covering the seeds with hot, almost boiling, water and leaving to cool. This will have to be repeated several times until some of the seeds start to swell. These plump seeds should be removed and planted, while the rest of the seeds receive more hot water treatment, until they in turn swell and take in water. There is very little chance of fungal contamination by this method.

Seeds should be planted either where they are to grow or planted in very long pots, at least 40-50 cm deep, as the first action of the small plant is to send down a deep tap root. Seedlings with two cotyledons and no true leaves often have tap roots 30 cm long. When still small the seedlings should be grafted or budded with scion from a known producing tree (female), always making sure that some male trees are around to ensure pollination. These can also be grafted. There are several methods for budding carobs, such as "flute budding" and "T budding". A good grafting book should be consulted. The best time for all carob grafting is spring.

Foliage of young trees must be protected from foraging animals as they find the leaves very tasty. Cattle, horses and goats will all kill young trees by eating both leaves and bark. Some trees may be attacked by carob moth (*Ectomyelois ceratoniae*) - the grubs feed on the contents of the pods. This problem continues because old pods are left on street trees and this allows the pests to survive over winter to infect the next year's pods. Removal of all pods minimises damage.§

Staple Mediterranean Food - The Fig

The common fig tree (*Ficus carica*) provides succulent but extremely perishable fruit that is valued both fresh and dried. The fig is indigenous to Persia, Asia Minor and Syria and is today cultivated in most warm and temperate climates.

Fig trees may be propagated by cuttings or by layering. Cuttings of about 20 cm in length should be taken from one year old wood during the late dormant period. Best results are achieved with cuttings that have a terminal bud. Fig trees may be obtained faster from layering. When rooted, these will form plants ready to bear fruit in the first or second year after planting. Fig trees grow rapidly and need little pruning - a worthwhile asset in a permaculture system. They are self fertile, so one tree is sufficient for fruiting. There are many different varieties of figs - from large purple ones with deep red insides to medium brown and small green ones with brown insides.

The fig produces two sets of shoots and two crops of fruit in a season after remaining leafless and dormant over winter. The first crop, usually insignificant, occurs early in summer and the large second crop in late summer. Botanically, the fig is a remarkable form of fruit. It is actually neither a fruit nor a flower, but a hollow fleshy receptacle enclosing a multitude of flowers which are never exposed but ripen their seeds inside. So the flowers are concealed within the body of the fruit. The edge of the pear-shaped receptacle curves inwards to form a nearly closed cavity bearing the numerous fertile and sterile flowers mingled on its surface. The male flowers are mostly in the upper part of the cavity and few in number.

As the fig ripens, the receptacle enlarges greatly and numerous one-seeded fruits become embedded in it. The figs are borne singly in the axils of the leaves.

Fig trees are small, rarely more than 6-7m high, with broad, rough, deciduous, deeply-lobed leaves. They live to a great age and grow in a wide range of soil types. They thrive in well-drained sand, but tolerate heavy soils, and need no watering in summer.

Figs are an exceedingly perishable fruit and very susceptible to fruit fly, which is why you rarely find them for sale in fruit shops. In hot weather they will last only a few days. Raw figs are delicious to eat but with a large producing tree you need a lot of people to consume the fruit as it comes! Figs are eaten after cutting off the stalk. Some people insist on peeling figs but this is unnecessary for all uses. If you were brought up on your Mum's home-made fig jam you will probably have had several jam-making sessions. Fig jam is absolutely delicious and certainly hard to come by unless you have a friend who makes some. It's really easy to make - all you need are some figs, lemons and sugar: 4 kilos of figs to 3 kilos of sugar and 2 lemons. The method is to wash the figs and cut them up roughly, removing the stalks. Then sprinkle two cups of sugar over the lot. This should be done soon after picking and definitely on the same day.

Then allow to stand overnight so that juice is drawn out of the figs. Next day add the juice of the lemons and the skins (not cut up). Boil until tender, add the rest of the sugar and stir frequently. Boil until a golden brown colour and turning to jelly. Remove lemon skin and bottle in sterile jars.

If you have extra figs experiment with fig pie. Cut up figs, stew them with a small quantity of sugar and water, cool them, then thicken the mixture by cooking with cornflour to make a pie-filling consistency. Put into a cold, crushed-biscuit-and-butter piecrust. This has proved to be a delicious cold summer desert on hot nights.

If you live in a fruitfly-prone area, you should practise strict hygiene. It is essential that you pick all fruit as it ripens and pick up all fallen fruit every day and use or destroy it. All infested fruit can be destroyed by boiling. If you boil up infested fruit and cool it, it is all right to then bury it. But it is very important not to bury untreated fruit as this only continues the life cycle, because the maggots or larvae leave the fruit and burrow into the soil. These, in turn, take about 10 days in warm conditions to mature and burrow out of the fruit into the soil. And so the cycle continues.

The stage at which organic gardeners can control the fruit fly is preventing the maggots reaching the ground and producing more flies. You must watch the tree daily and remove ripe fruit and pick up and dispose of any that has fallen. Chooks under a fig tree make life easier, as they eat the figs (and maggots) as they fall. Also, when making jam you need a fair number of half-ripe figs as well as ripe ones, and this will give you 2-3 days rest before more figs ripen.

When you start to feel this is all too much of a bind - the fig season lasts about 4-6 weeks - don't forget that friends and neighbours without fig trees might appreciate some too.§

Fodder Trees - Sustainable Integrated Species

There is always a risk, when discussing the suitability of fodder species, that someone will decide that a particular species sounds ideal and sets about planting a broad scale monoculture. Should these notes fall into the hands of novices to permaculture, I shall emphasise that this kind of practice is not the most efficient or sustainable approach, does not promise a balanced year-round diet and provides no diversity against pests or the unpredictable weather you may experience. In short, it is not permaculture.

The design of the forage system to suit the landscape, type(s) of animals, species of plants, type of fencing and so on, is, of course, the key to success. If the animal can harvest its own food without destroying the system, then you are left with more time to plant trees. Fenced hedges of tagasaste, coprosma, willows, poplars, pampas grass, banna grass, elephant grass and occasional oak, carob, honey locust and chestnut, all mixed together, are good value for cattle, sheep, goats, horses, donkeys and even pigs.

The hedges are also best sown with medicinal herbs such as anise, asparagus, beet, balm, borage, chicory, comfrey, cress, daisy, dandelion, dill, dock, fennel, foxglove, garlic, hollyhock, kale, lavender, lucerne, mallows, marjoram, mint, nettle, parsley, plantain, rosemary, rue, sage, southern-wood, sorrel, tansy, thistle, thyme, valerian and violet. Medicinal trees include alder, ash, beech, birch, elder, elm, hawthorn, hazel, holly, juniper, lemon, lime, quince, raspberry, senna, walnut (leaves are good buttermilk food) and witch-hazel. Stock will clean everything back as far as they can reach through the fence, but will only browse the herbs as they require.

Fences need to be strong. Consider a front line of 'live barbwire', such as *Rosa rugosa* or loganberries. *Rugosa* hips contain high vitamin C for humans living in a polluted world.

The fodder hedges are best placed to create many small paddocks, containing as many grasses, including pampas, banna and elephant, clovers and lucernes, as you can pack in. Ideally, hedges should run off a central water trough so that changing paddocks becomes as easy as opening and closing gates at drinking time. Allow paddocks long recovery time after grazing or browsing. This reduces the incidence of rampant weeds, which usually only get a stronghold because all their competition is being eaten. Stock should only stay in each paddock for a few weeks each year.

The increased shelter and nutrients provided to these small paddocks will also add to their productivity. The winter paddocks are the place for oaks and carobs, spring for mulberries (especially for pigs) and in summer moist paddocks will grow willows, poplars and bamboos.

There is no need to fall into the square paddock syndrome just because your property is that shape. Follow contours, create wind breaks or even wind funnels, if you are relying on wind power to pump water or power electric fences. Remember, the more edge you are creating, the more access the stock have to the fodder belts. "And the more bloody strainer posts," I can hear the practical farmer say. My only defence is that it will be worth the extra effort. Besides, smaller runs of fences don't need such heavy strainers.

What we are effectively doing is stepping a synthetic forest into grassland. Following the patterns that Nature shows us will greatly assist. When Nature steps forests out into grasslands she sends pioneers to pave the way.

Often these pioneers are totally displaced by the succeeding forest and generally are only found on the edge (hedge). Such a species is tagasaste. Be cautious of the yellow flowered broom which is poisonous to stock. Folks living in the tropics can use leucaena instead of tagasaste. Leucaena will even grow in Mediterranean areas. There is also tree medic (*Medicago arborea*) which is a useful fodder shrub. The tagasaste, tree medic and leucaena are all nitrogen fixers, their foliage is high protein stock food and incidentally, makes wonderful mulch. They play the same role in your hedge rows as the clovers and lucernes play in pasture - nature's nitrogen fertiliser factory.

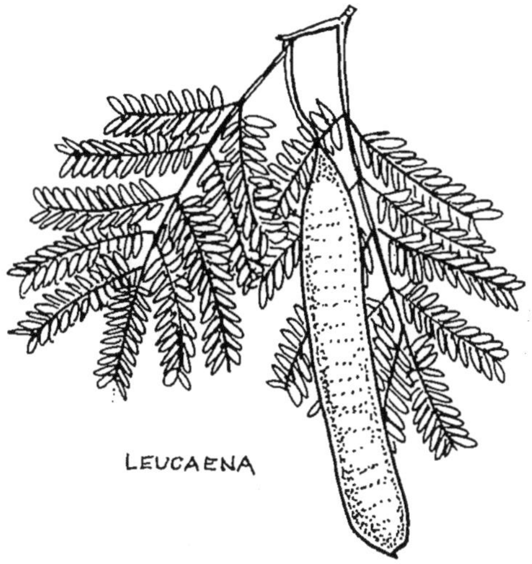

Tagasaste is extremely fast growing and although only short lived (usually 5-10 years but occasionally to 50 plus years), will reseed rather prolifically. It is also an excellent winter and spring bee forage. Its potential to 'overwinter' hives should not be ignored. Commercial beekeepers are spending a lot of fossil fuel taking hives north for winter. The seed is also a good chicken and pigeon forage. So some of your hedgerows can also provide chook forage, as can mulberries, coprosma and cotoneaster.

Coprosma is dioecious. You need some males but mostly females for optimum berry production. Most stock eat the leaves, so unless you are feeding poultry or collecting seed, you aren't chasing maximum berry crop. It is good to have them self seeding, so go for both sexes, and try to get either seedlings or cuttings from many different trees so you have genetic diversity and all your plants don't have the same weaknesses, such as susceptibility to disease and pests.

Carobs are also dioecious, so plant them in close association, a half dozen or so at a time to ensure males and females. They are best planted out only a couple of weeks after germination (from 1 litre milk cartons.) Much longer and the tap root becomes turned up - not good for dry areas. We lost all our carobs in one planting trial in the central wheat belt, simply because our potting soil contained some sawdust. White ants in those parts are starved for wood. They came for the sawdust and took the carobs as well.

Probably the best pioneers for drier areas are the native acacias. Mulgas and jams come to mind but there are plenty of others.

The best of the oaks for the wheatbelt and drier areas is probably the cork oak. As an evergreen it can do its growing over winter when there is plenty of water. They are extremely drought resistant. We plant the soaked acorns (acorns must be fresh) directly into prepared holes in early winter. The root can go down 600 mm before any leaves appear.

Acorns are first grade high energy stock feed, 80% fat and carbohydrate, which fall to the ground in autumn and winter just when the stock need the extra warmth. Cork oaks are also the source of commercial cork. The better cork is produced on the poorer soil and in drier conditions.

Honey locusts, bamboos and poplars all tend to send up suckers. Any that go beyond the fences will soon be devoured.

The hedgerows make wonderful wildlife corridors. A narrow hedge, 2 - 3m, is appropriate on either side of a road. You may as well include walk paths or even vehicle access through some of your hedges. Most of the herbs are also valuable human medicines, and carobs, honey locusts and acorns have all been staple foods of various peoples throughout the world.

Woodlots are also good short term forage systems if set out correctly. Be careful not to exclude grazers completely from pasture. If you are starting with a grazing property, and not wanting to lose much feed during establishment time, then the logical way to go is slowly first. Get your design clear then implement it over five or so years. As the first hedges become productive (second year for tagasaste) they will more than compensate for the lost grazing space of further developments.

Deep ripping before planting hedgerows will prove worthwhile on most farmlands. The best time to rip is the driest part of summer to really crack heavy claypans. It will require a heavy duty machine, but you will notice the difference in growth of plants.

These are just a few ways that you may wish to go. Whatever you decide, careful planning is the first and most important step. Remember, make your hedges diverse, as many species as you can arrange, all growing together, just like in a natural forest, and go for clumps of trees rather than long lines. Think of all the things that you don't want travelling down your hedges e.g. insect pests, disease and fire. They will all find it harder going if they don't have long lines of their particular host plant to travel along.§

Alley Cropping

Introduction

One of the characteristics of a Permaculture system is its polycultural nature, where a mixture of many different species live together. It is modelled on Nature where polycultures of great diversity enable species to exist and evolve as long as the sun continues to shine.

Alternative agricultural systems for both the urban and rural landscapes are needed. One recent development has been a technique called "alley cropping".

Alley or hedgerow cropping consists of plants, trees or shrubs grown in wide rows. A crop, such as cereal, vegetables or fruit, is grown in the inter space of the thus created "alley".

Alley cropping offers benefits to the grower and the environment. It deserves to be established in farming areas because it can go a long way to modifying them.

Benefits

Alleys, in much the same way as your backyard, can be very creative and productive places. Some of the benefits include:

- Increased yield, due to the increased organic matter, and nutrients supplied to the soil and the crop.
- Use of deep-seated nutrients, and moisture. The tree and shrub roots will seek out and find minerals and water held deep in the soil. These are recycled through the prunings of the trees or shrubs on each side of the crop. The crop will then benefit from this recycled organic matter deposited in the topsoil.
- Improved soil structure. The increased soil organic matter will improve the soil structure. The flora and fauna of the soil will increase and their actions will cause the improvement of soil structure, e.g. the benefits of earthworms.
- Prevention of soil erosion. Particularly on sloping ground the tree/shrub lines on a contour will physically stop and catch the flow of water and soil down the slope.
- Multi-functional effect. Alley cropping will provide additional products, such as forage for animals, composting material, timber, firewood, fruit and nuts.

Disadvantages

There are a few possible disadvantages that can occur, some of which include:

- Competition between tree and crops for light, moisture and nutrients.
- Reduction in the area available to the crops.
- Labour and cost of tree planting, pruning of the trees/shrubs and distribution of the prunings/mulch.
- Limitations or restrictions in land use.

These possible disadvantages are outweighed by the indicated benefits. Most of these disadvantages can be easily eliminated by design planning and management practices.

The need for more labour can be a positive effect in a depopulated rural landscape.

Applications

For coastal farming areas alley cropping could be most beneficial. The trees or shrubs which form the alley can be pruned in the autumn. This will allow winter sun to reach the crop. During the summer the crop will receive shade which will reduce water loss and exposure to excess ultra-violet light. Both these can affect the growth and health of the crops.

In large-scale use, trees can replace shrubs. Agro-forestry and the inclusion of animals can be a form of alley cropping. The windbreak effect, particularly during summer when the hot, dry winds blow, is another important benefit from alley cropping.

Plant Selection

Plant species should be selected as the alley border trees and shrubs if they possess most of the following characteristics:

- Ease of establishment and rapid growth.
- Good coppicing ability (regrowth after cutting).
- Nitrogen fixing capacity.
- Deep-rooted and different root distribution to the crop.
- Multiple use - such as forage, firewood, erosion control.
- Ability to withstand environmental stress such as drought, waterlogging, extremes of soil pH.
- High leaf-stem ratio.
- Small leaves and leaflets.
- Resistance to pest and disease attack.

Plants which do possess many of these features include species of *Acacia, Casuarina, Medicago, Alnus, Cassia, Leucaena* and *Sesbania.*

Some suitable crops for alley cropping include cereals and many vegetables. The practice of using these types of plants could dramatically improve the health and increase the productivity of agricultural soils.

Conclusion

Alley cropping, which originated in Africa, could provide real benefits to agriculture both in the urban and rural landscape.

It can be scaled up and down. For example, in the urban situation shrubs can replace trees. Differing crops, such as cereals, fruit, nuts and vegetables, can be grown in either situation. It seems to be a very suitable technique for use in permaculture design.

It would be important where alley cropping is established to keep records and make regular observations.

These would be of great value to others who might wish to give it a trial. Is alley cropping up your alley?§

Tagasaste - The Tree for Every Permaculture Backyard and Farm

Tagasaste (*Chamaecytisus palmensis*) is an attractive, evergreen, leguminous tree or shrub, that produces high protein fodder, grows quickly and easily in a wide range of climatic conditions, (tolerates frost but doesn't tolerate waterlogging), and is relished by stock.

Tagasaste as a forage tree can triple the carrying capacity of farmland: yield increases in crop and pasture occur when grown in combination with tagasaste. Tagasaste can also be used in erosion control, and it has fire retardant properties. It is also an excellent and economical bee forage plant, as it flowers profusely and over an extended period, depending on times of pruning.

For permaculture backyards

- This tree is a superb windbreak, and will spread and cover a 5 m x 5 m area, but can be pruned and trimmed to any shape - even a hedge, which, in turn, is a fire retardant if pruned from an early age. It doesn't like being cut back on big wood.
- It is excellent as a nursemaid and windbreak for delicate trees such as tropical fruit trees, as it is a "nurse tree" and nitrogen fixer. In such a situation it should be pruned so that it shelters, but doesn't overwhelm, the delicate tree, and the prunings can be used as a most effective mulch - either stems and all, or else put through a "garden muncher" first.
- This "muncher" mulch is also really good for wormfarm feeding, mixed with horse manure or whatever else you usually use.
- Your rabbits will love the leaves and stems, and just cutting for them every day will keep your tree well pruned.
- The leaves can also be mixed with poultry food, and of course the seed is loved by chooks. The tagasaste is a real chookyard essential.
- It is an excellent bee forage plant, and provides a light and delicious honey.
- On top of all this, the wood is an excellent firewood - even the thin sticks - but be warned, cut it into fireplace sized pieces as soon as possible because the wood gets very hard if left to season. The trees are not always long-lived, so as they die, take them out and cut up the lot for firewood.

Altogether a 'must' for every backyard. Note: As tagasaste won't stand waterlogging, a raised bed is essential in that sort of situation, and it does need to be well fertilised, so a good manure mulch is needed at all times, plus the application of trace elements at planting time if in a deficient soil.

If you ask your nurseryman for 'tree lucerne' instead of tagasaste, watch out that you are not given the yellow flowered lucerne tree, which can be a pest in bush areas and also is a poison suspect. Tagasaste is a white flowered tree or shrub.

For broadacre areas

- Direct seeding is best for large areas, using scarified seed which will give an 80% to 90% germination. Inoculants are not always essential because there are so many nitrogen fixers in our bush, but cowpea or lotus major inoculants can be used quite successfully.
- Seedlings for smaller areas need to be got in at the break of season to allow roots to get down before rains cut out.
- Deep ripping is costly and not essential except in perhaps claypan areas.
- While you can transplant quite large seedlings, perhaps small seedlings, with minimum root damage, and therefore potential drought proofing, would be better in dry areas.
- Useless land, such as rocky hills, are fine as sites, but look for well-drained areas to avoid waterlogging.
- The usual spacing is 2 metres between trees and either 4 metres (1250 trees/ha) or 5 metres (1000 trees/ha) between rows, depending on your tree harvesting machinery. Wider spaced rows are needed if cropping between rows to increase your crop yields. Hay can be grown and cut between the rows in the establishment stages. The four metres between rows, with pasture between, is ideal for a 'stock help yourself' basis feeding. In this case, the young trees need to be pinched out early in order to make them multiple-shoot just above ground level, and to also prevent ringbarking of trees by 'self helping' stock. Naturally, stock need to be monitored carefully, because if left in for months till short of feed, they'll attack the trees.
- Fertilising is important. As a rule, trace elements (for example, copper, zinc and molybdenum) need to be applied at 200 kg/ha at establishment of plot, and top-dressed thereafter with an annual 200 kg/ha of super/potash 5 to 1. Alternatively, mulch heavily a few times a year.
- Established plots are everlasting anyway because seedlings are always germinating. To encourage this, leave some trees to produce flowers but even these can be pruned twice a year and still flower.
- Plantings can be done on a whole paddock basis, with room to crop and/or cut hay between rows if required. Animals can be managed on a 'self help' basis, al though treetops may still need to be lopped with either a tractor and anything from a modified mower to the special lopping machinery now available, or with long handled parrot jaw loppers for smaller scale plantings. As long as the small trees are pruned to encourage multiple low trunks, cattle in these paddocks can 'self feed'.

Stock must be managed - watched for overgrazing - and about 4 months left between grazings to allow trees to reshoot. In winter, pasture is usually preferred anyway. These paddocks are great feed and shelter for lambing ewes.

- In the establishment years, hay can be grown and cut between the rows. So the lean summer/autumn periods are catered for from the time you implement a wide scale fodder tree planting program.
- Thick hedgerow plantings do have a place in the overall farm plan, and they are a fire retardant as well.
- Good fencing is necessary in the early stages. Rabbits can be a problem, and tyre tree guards are a suggested temporary solution. Grasshoppers can destroy the trees when small, but flocks of guinea fowl will take care of that.
- If cutting and feeding out, any sticks left over are easily burnt in your stove.
- Trees will grow by cuttings if you find you have a superior tree which warrants perpetuating.§

Tagasaste - or White Flowered Lucerne Tree

Tagasaste had its known origin on Las Palmas, one of the Canary Islands. All the islands in this group have a low rainfall, but on Las Palmas only, the crops grew well and the stock were thriving. One astute observer noticed this, but also noticed that a certain shrubby tree grew abundantly and was cut and used as fodder for the stock. This tree was tagasaste. Some seeds were collected and taken to Kew Gardens in London, and it was from there some seeds were sent to the Colonies.

Tagasaste is an excellent shelter tree and assists in preventing soil erosion. Grazing animals seem to make better use of available dry fodder when fed even small amounts of green feed (this can even be young eucalypt growth). Crops grow better between rows of tagasaste and there is a 20% increase in pasture growth in the shelter of these trees.

Being a legume, it has the power to use free nitrogen from the air and store it in root nodules. However, unlike clover, it does not seem to require inoculation, apparently being able to use the rhizobia occurring naturally in the soil. It flowers profusely, providing an excellent source of bee food.

It also seems to be fire resistant. On Kangaroo Island, off the coast of South Australia, some years ago a raging bushfire so threatened to engulf the homestead on a property that fire-fighters had to abandon their attempts to save the home.

When the fire had subsided and they returned to inspect the damage, the homestead, complete with its surrounding tagasaste hedge, was the only area left unburnt.

Untreated seed can be very difficult to germinate. One way is to bring water to boiling point, remove from heat, then drop the seeds in. Leave to cool overnight, then plant, barely covering the seeds with soil. This should give 75% germination.

It is now possible to buy scarified seed, that is, with a nipped end, and these should give about 98% germination. This has made large scale plantings feasible because the scarified seeds can be direct seeded in rows and fertiliser incorporated at the same time. Clay soils should first be deep ripped.

When used as a browse shrub there would be no need to use herbicides to control weeds as the animals will do this. However, if overstocked they can kill tagasaste if they start eating the bark. One answer to this is to have a lot of trees and fewer animals. It was noticed also that there was little or no inclination on the part of the sheep to eat the bark when they were also supplied with mineral supplement licks. In addition, a system of trimming right back to half-a-metre from the ground, thus encouraging multiple growth near the ground, tended to overcome the problem of stock damage to the tree.

Finally, tagasaste, which is frost tolerant, must surely rate very serious consideration as a forage tree as a complement to the normal system of pasture feeding.§

The Mighty Oaks

The genus Quercus (the oaks) contains over 450 trees native to areas as far apart as North America, temperate and subtropical Eurasia, North Africa, and South America. Almost all species have excellent, durable timber, and many have been used in wooden ships. However, as well as being magnificent large trees producing excellent timber, oaks have other useful products including: cork, from *Q. suber* (in fact, the bark of this oak); tannin from the bark of several species; ink from the insect leaf galls of some old world species; and edible acorns, for both animals and people, from some species.

Many oaks are native to North America, where the American Indians used the acorns of several species for food. As raw acorns almost always have too much tannin for direct consumption, the Indians hulled the acorns, ground them into meal, then washed this with hot water to remove the tannins.

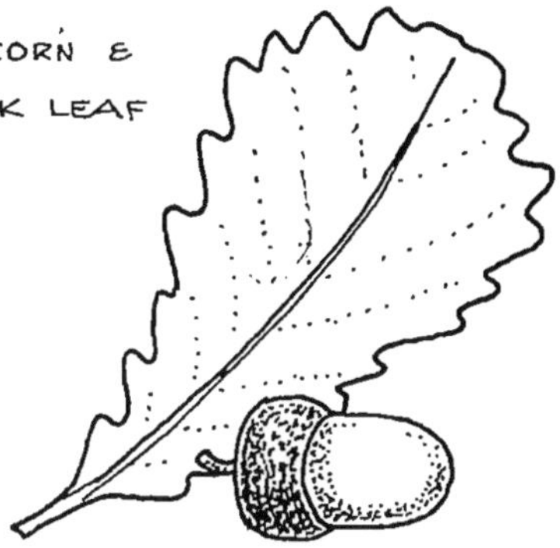

The flour which resulted was used to thicken soup or to make bread which was tasty and would keep without spoiling. Roasted, ground acorns have often been used as a coffee substitute.

Oaks, which are mainly deciduous (not all oaks are) are very beautiful and have often been used to line the streets of cities. Some common oak species include:

- *Q. suber*, the cork oak, from the Mediterranean regions, which produces large acorns, most edible roasted. Raw acorns are good pig food. Cork is also harvested from this oak and has a wide variety of uses.
- *Q. robur*, the English oak, from Europe, West Asia, and North Africa, which produces acorns which are good roasted or ground into coffee and good raw for pig food. Acetic acid and tannic acid can be extracted from the bark and wood.
- *Q. palustris,* the Pin oak, has lovely little acorns that look like chocolate drops, but these are not listed as edible.
- *Q. ilex*, the Holly or Holm oak, from the Mediterranean regions, which produces sweet edible acorns.
- Q. *lusitanica,* the Portuguese oak, from Spain and Portugal, is a semi-evergreen oak, which is similar-looking to the English oak.§

Composting in a Permaculture Garden

Introduction

In permaculture design, composting methods are usually incorporated in sheet mulch garden beds, paths and via the actions of insects and animals.

However, there is a limited place for composting in permaculture. To do it well takes considerable skill and large amounts of materials.

Composting methods

Four methods of composting are common.

1. Rough composting. A square-based pile of plant material and animal manure on which surplus and spent plant material is heaped. This is a slow method. The pile will be broken down and used in other composting methods, such as the Indore and Sheet Mulch garden beds.

The walls of this pile are constructed out of folded newspapers, which gives a use for your old newspapers and insulation of the pile from the cold. Temperatures of 40°C - 60°C are required to reduce pathogens including weed seed.

2. Indore composting method. A method devised by Sir Albert Howard while working in India. Layers of green plant material, animal manure and soil are built up to two metres high, and the pile is covered with a weed-free organic material. This is necessary to maintain high temperatures needed to complete the composting process.

When constructed in piles or heaps, it is difficult to maintain straight sides, and this does not make for maximum use of space. Pits or above ground bins provide for better use of a given space; they can be made of various materials, such as recycled pallets which make quick and convenient compost bins.

Covers are essential to maintain moisture and prevent water-logging during winter. Water content is an important factor in producing finished compost and humus. Three bins make for an efficient method of composting. The first one is used for bulk material, such as green plant material. The second is where a complete Indore method is constructed.

For quick results, temperature and water content are maintained to the correct levels. When the high temperature falls and the compost drops in height it is time to turn it over into number three bin. This is ready for use. A cover needs to be provided to conserve it as it is being used.

As you can see from the above, composting can consume vast amounts of materials and energy. There is energy lost while it goes through the composting process.

3. Sheet Mulch garden beds. Sheet mulching is no doubt understood and used by many practising permaculture people. It is well covered in the permaculture literature. It has many advantages over the two previous methods mentioned. The advantages are:

- The composting occurs in situ (in the desired place).
- Recycling of plant and animal wastes occurs.
- Planting occurs immediately after the sheet mulch is laid.
- It is done only as required.
- Weed control occurs as a result of it.

It should be noted that sheet mulching is usually only carried out in zone I of permaculture design. Vegetables are mainly planted in this sheet mulch. Herbs and flowers are included for companion planting purposes. This is done to maintain their health and pest resistance.

There are two other composting methods being carried out in the Permaculture garden. Both have similar characteristics, although their scale differs.

4. Jean Pain Method. A pile of cuttings and prunings is made. Nothing is added to the heap. No manure, soil or herb activators.

Naturally present bacteria and an abundance of water, (the heaps are soaked in water after construction), is all that is required.

Soon, an abundance of humus, black sticky matter, is present. This can be screened out and used for the sheet mulch garden bed construction. The larger pieces can be used for paths. Paths provide an informal composting system, comprising shredded brushwood and prunings. These paths are covered with a liberal dressing of brushwood or prunings each year. It quickly compacts and provides a pleasant surface. Plants growing either side of the path gain nutrients from the breakdown and composting of the path. Earthworm activity is high in these paths, which produces nutrients in the form of their casts.

Paths in permaculture provide many functions as well as their primary role of providing access - an example of permaculture principles in action.

Plants that exhibit rampant behaviour or are going to seed in an unwanted situation can be controlled by regular slashing prior to bolting and seeding.

Their slashed portions provide material for composting in situ or sheet mulching garden beds.

The slashed plant continues to grow, and its root growth adds organic matter to the soil. Life in the soil, in particular worms, is high due to the abundance of organic matter descending from above.

Conclusion

Composting in the permaculture garden occurs primarily in situ, with sheet mulch garden beds in zone I and other gardens and animal products in zones II and III. Large, energy-wasteful and, in many cases, unsightly piles and heaps of rotting plant matter are not necessary. Permaculture as a self-sustainable system requires our intellect, not our blood, sweat and tears.§

Mulching

Some useful mulches include:

Pine Needles. Very helpful in holding moisture, repelling bugs and especially reducing alkalinity (may be a problem if you have your own bore).

Grass Clippings. Heat up to a very high temperature in early stages. Can kill weeds if spread thickly and sprinkled liberally with poultry manure then topped with good compost or sifted garden soil.

Can promote early growth if used as a seed bed in a cool spring. Also good in seed boxes.

Seaweed. Probably the best of all natural mulches. Holds moisture, supplies essential minerals and trace elements and has disease inhibiting qualities.

Stones. Heavy to handle, rough on hands and expensive if you have to purchase them. However, they keep the soil warm in winter, cool and moist in summer (also a refuge for worms). Good also for low maintenance parts of the garden.

Plastic Sheeting. Nasty stuff to handle, but does keep in the warmth if needed and kills weeds when weighed down over weedy paths.

Wood and Bark Chips. Very good. Once they are thoroughly wet they will settle down and soon be teeming with worms. Provides some nutrients and conserves moisture.

Shredded Paper. Best if soaked first. Very good as a 'starter' in water repellent sand.

Leaves and Twigs. Very effective. All shrub and tree leaves are raked up to the drip line without any other mulch and plants always have healthy dark green foliage and green shoots. Any spare leaves are used in compost or as mulch for newly planted subjects.

Straw. Great! Stays very damp for several days in heat-wave conditions. Lasts very well, and is devoured by earthworms, leaving a bonus rich supply of humus and worm casts. Unfortunately, it attracts slugs which multiply rapidly.

Sawdust. Really needs composting before use. Heats up when thoroughly wet, so is good for seed germinating boxes if covered by an old sack then a good seed mixture. Once aged, it is a good mulch as, along with bark or woodchips, it attracts earthworms and very gradually disintegrates into a soft crumbly soil.

Ageing involves soaking with water several times, turning and allowing to dry out between treatments. It is helpful to mix sawdust with the poultry manure at first treatment, as this resolves the problem of nitrogen loss and all the weed seeds are killed in the early heating up process. Apart from mulching, this mixture can make a good 'no-dig' garden if mixed also with other composting materials, such as comfrey leaves, chopped lemon skins, and bone dust. It needs to be enclosed for stability and lasts for a long time.§

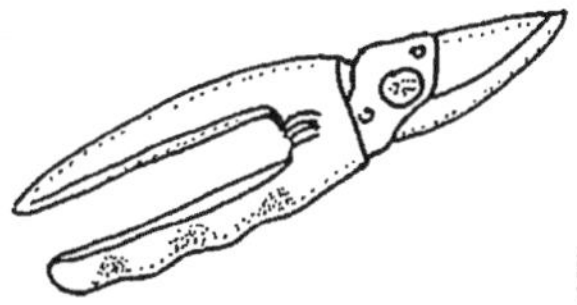

Pruning

Everyone can deal intuitively with fertilising, watering and pest control, but not with pruning. This has come about because of the extraordinarily complicated systems developed in the past. However, personally I think more nonsense has been written about pruning than any other subject.

Once I was discussing pruning with a very good permaculturalist, and after about half an hour's mucking around with one small tree we thought we had it about right, with all the buds and shoots just where they should be. Then an old non-permacultural ex-professional pruner came along. So we moved to the next tree, handed him the pruners, and asked him to show us how it was done. He looked at the tree for a few seconds and in 30 seconds it was all over. There was no mucking around with the finer points - he just whipped off all the right bits. Now I emulate his style.

There are only two basic factors. You are pruning to get more of the right fruit, and to get the tree into the shape that is convenient. More specifically, I prune with the following ideas in mind:

- Prune to keep the tree at a size and shape that is convenient for picking, pest control, shade, aesthetics or whatever.
- Prune to keep the growth down as excessively large trees need more water and fertiliser.
- Prune to keep the fruit low down or at least not too high, unless you like climbing ladders.
- Mercilessly prune to get rid of errors (dead wood, crossed branches, previous errors, diseased bits and so on).
- Prune to keep the density of foliage and fruit even and just right to let in light. This means avoiding having all the foliage at the top and therefore no growth on the bottom shoots - otherwise the tree will just get taller every year.
- Prune to force the growth into fruit rather than branches, unless you want the tree to grow.

Other important considerations are:

- Don't restrict yourself to pruning branches. If there is an excessive fruit set, then prune off lots. It is better to have 200 large fruit than 500 little ones.
- Don't regard pruning as primarily a winter activity. I prune mostly in the summer. In winter, it is nice to do a bit of touching up when you can see all the branches, and that's the time for really major surgery like setting up a whole new branch.
- Don't make large cuts unless you want new shoots. If you make big cuts, you get big new shoots. This is a major advantage of summer pruning. Always remember to cut off enough of the new shoots, or you will get too many.

- Don't delay with young trees. Buy only trees with the shape you want, and make sure they stay that way. With young trees this means regular and careful pruning.

  Bending is equally important - that is, bend branches to where you want them. I use bricks tied with loops of baling string, and as many as are necessary. Four on a young tree is quite normal.

  This is particularly useful in spreading branches in nursery stock which has been crowded and grown straight up. A bit of judicious bending will lay down a good shape for years to come.
- Don't read books about complicated formal pruning styles unless you happen to like that sort of thing.

A. TREE SHAPE

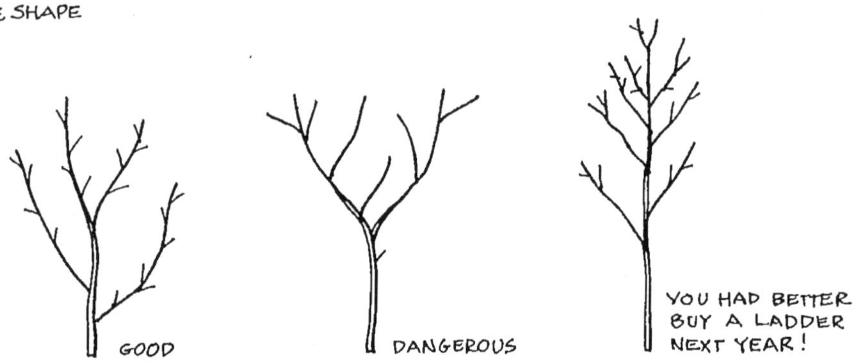

B. SINGLE BRANCHES

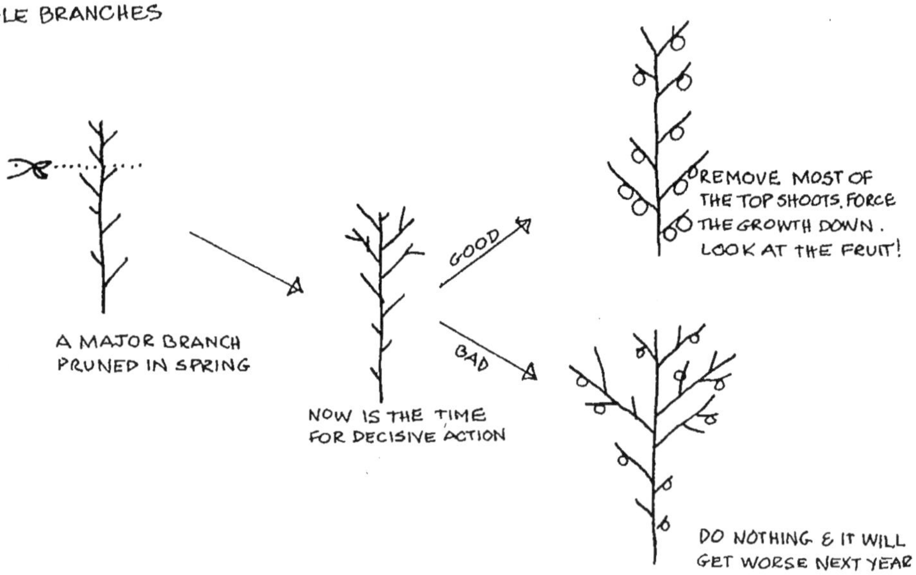

The other side to pruning is to think in terms of pruning off whole trees. This has its basis in modern commercial intensive plantings (such as trellises). The idea is that you plant at extremely high densities and get a high yield per area in the first years when the trees are small.

After a while you can prune back severely or even prune out whole trees. This is particularly useful if you put in several different varieties, because you can eliminate any trees that don't do well or if you don't like a particular variety for some reason.

When you are getting started, the idea of killing a tree sounds like murder, but once you get 500 plums from each of four varieties it begins to look like desirable euthanasia.

Finally, a few specifics and practicalities: if you are spending more than a few minutes on a small tree at one time, then you are doing too much. Most of the few minutes should be spent looking at the tree, followed by a burst of strategic cutting.

Citrus and most evergreens (like the guava, avocado and custard apple) are pruned mostly for shape. This means chop off the top to keep the height down and chop off any branches that stick out sideways and get in the way.

Peaches and nectarines fruit on last year's wood, so you have to thin out the laterals by cutting some off. This pushes growth into the remaining laterals and fruit, and promotes new laterals for next year. Prune the top and sides to keep the height and shape right. Keep the number of fruit down by reducing the number per cluster as well as the number of clusters. For example, you could reduce all multiples to singles and remove half the singles in a good year.

Apricots fruit on wood a few years old, so pruning is really only for shape. Apricot trees get very large and hard to pick, so try to keep them down. Then thin the crop without hesitation if the crop is big.

Plums fruit on those little laterals along the branches so you prune the tops of the branches to force the growth into these laterals and the fruit. Fruit thinning is often required, and try to let the tree grow in size only a bit each year Grapevines are quite different. Initially, prune to set up some wood as a skeleton on the trellis.

After that, pruning is very simple - for every two annual shoots, remove one completely and leave two buds on the other.

A few varieties are different (including sultanas) and will not fruit if pruned this way. Consult a book.

Figs and mulberries don't really need to be pruned, so you just keep cutting them back to keep the size down. They are very forgiving and can be pruned into almost any convenient shape.

Above all, don't get too obsessed with pruning - most trees fruit pretty well anyway!§

Seed Saving

Choice of plants

In seed saving, the basic rule is "Never save seeds from hybrid plants or allow your plants to cross pollinate with hybrids, as the seed may be sterile or revert to one or other of the parent plants".

Your choice of plants to grow seed from will determine the kind of vegetables you will grow in the future. Choose a couple of vigorously growing plants, which are resistant to insects, for their ability to grow well in your environment, in the colder districts, in the hotter regions. Those plants which go to seed early in the season are of no use, as their progeny will also have this trait. Another selection may be for the size of fruits or for early bearing characteristics. You may need special qualities such as lack of thorns or strings - as in beans. All these distinguishing features will carry forward into the next generation of seed. Having selected the plants you want to grow seeds from, tag them and the individual fruits, so that other members of your household will not pick them for food.

Storing the seeds

Most authorities advocate the use of airtight containers for storing seeds. However, seeds breathe at a very reduced rate, and if storing seeds for any length of time, (over one year) it is recommended that the tops of the containers be opened for a short while to replenish the oxygen supply. The use of airtight containers is essential to keep out humidity, which could allow the seeds to go mouldy or even to start to germinate in extreme cases. Manila envelopes, if kept in dry places, are adequate for general use. Keep seeds in a dark and cool spot. Do not use plastic bags for storage.

When drying seeds, do not leave them in the sun - they will become completely dehydrated and die. An airy shed or garage is ideal, or any dry place with a current of air. Remember, dampness is the worst enemy of stored seeds.

To prevent spoilage by moths and other insects which may penetrate the containers in which the seeds are stored, use bay leaves and diatomaceous earth to mix in with the seeds. (Diatomite can be provided by swimming pool suppliers.) Viability of seeds depends more on a cool dry storage place than anything else.

Seeds can also be successfully frozen, as long as they are dry, and indeed, sometimes freezing can increase the vigour of the resulting crop. Experiment!

Packs for freezing will have to remain completely sealed against moisture getting in and freezing conditions completely desiccating the seed. For a short period it may be an ideal alternative.

Pollination

There are four main means of pollination: a) insect pollination; b) airborne (wind) pollination; c) self pollination; and d) hand pollination.

a) Insect pollination is the most common, and occurs in plants which have attractive coloured flowers, or odour, or have nectar or excess pollen.

All plants which are insect pollinated will obviously be prone to cross pollination.

b) Airborne or wind pollinated plants, such as corn, many trees, grasses, and beetroot, still need to be separated by variety, otherwise pollens will get mixed and the seed will vary from the parent.

c) Self pollinated plants such as peas, beans and tomatoes can also be cross-fertilised as insects can bring pollen. However, the chances are smaller than in the insect pollinated varieties.

If you want to keep your seeds true to type, the developing seed heads can be protected against the occasional insect by covering it with a paper bag until the seeds are formed.

d) To hand pollinate plants, pick the male flower, which is the one without the swelling at the base, and press the pollen it contains on to the stigma of the female flower. This is the one with the miniature ovary or fruit at the base.

In this way you will be sure that seeds will be formed, and in most cases, more seeds will result than if left to the bees. This is especially true of many vegetables such as the New Guinea bean and the luffa.

Saving your vegetables

The following are seed saving tips for individual species:

• Beetroot and Silverbeet - biennial, airborne pollinated.

Keep perfect beets over winter. Keep stalk supported by stakes as plant becomes over a metre high when bearing seed. It has numerous branches. When about two-thirds of the seeds have become dry, cut the stalks and store in a bag in a dry place. Thresh after two or three days, when seeds will drop out. If left longer, the dry stalks will mix with the seeds. The seed is always ripe before the stalks dry.

• Carrot and Parsnip - biennial, insect pollinated.

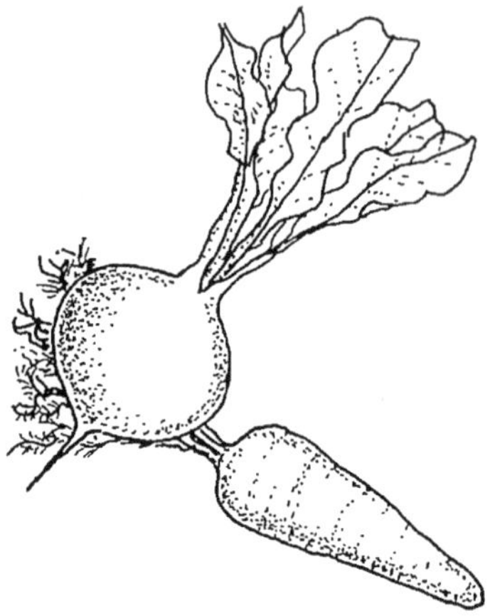

Also grows over 1 metre high. Whole head can be enveloped in a large paper bag, or individual branches cut off when beginning to dry. Parsnip seeds are viable for only six months at the most.

• Brassicas - Chinese Cabbage, Turnip*, Collard*, Cauliflower, Brussel Sprouts*, Kohlrabi*, Broccoli, Radish and Kale*. Note: * = biennial.

Most of the above plants will readily cross with one another, so grow one variety only at one time, or isolate by at least one kilometre. Cut all branches as they become dry and store in a dry place. If left on the plant, seeds tend to shrivel up. Thresh all together and dry seeds thoroughly, then store. All these plants are pollinated by insects.

• Lettuce - self pollinated.

Select late-to-seed plants. When seedhead starts to form, strip lower leaves and allow plenty of air to reach all parts of the plant. No isolation distance is needed between plants so you can allow many varieties to go to seed at the same time.

Flowering heads are tall so staking may be necessary. When the seeds are ripe they will be covered by a white fluff. Cut each stalk as it shows this fluff, as many seeds can be lost to birds or by being blown away. Store in a dry spot and thresh all together. Store only when quite dry. Leave the stalks on the ground and any left-over seeds will germinate underneath, away from doves and other birds which love them.

Note: To make lettuce seed germinate in hot summers, either sprout the seed in the refrigerator or leave the seed in there for 24 hours and it will then germinate. The cold stimulates it into growth.

Some herb seeds, such as chamomile, also need stratification; that is, being stored in the refrigerator for up to 3 months.

• Onions - biennial, insect pollinated.

Seed is ripe when upper part of stalk and seed pods are becoming yellow. The seeds are black. If the head is covered with a bag in the usual way, the seeds could go mouldy or rot. Cut the head and allow to dry before threshing into a large bag. Seeds which remain can be removed by washing. Place the partly emptied head in water, stir well, then the seeds should sink. Pour off the water and chaff and dry seeds well on paper or cloth.

• Peas and Beans - self pollinated.

Leave selected early pods on the vines until dry. Shell by hand when pods are crisp.

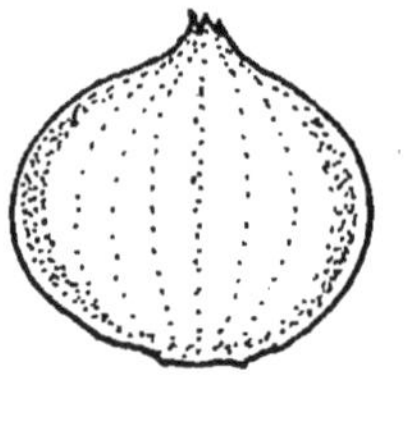

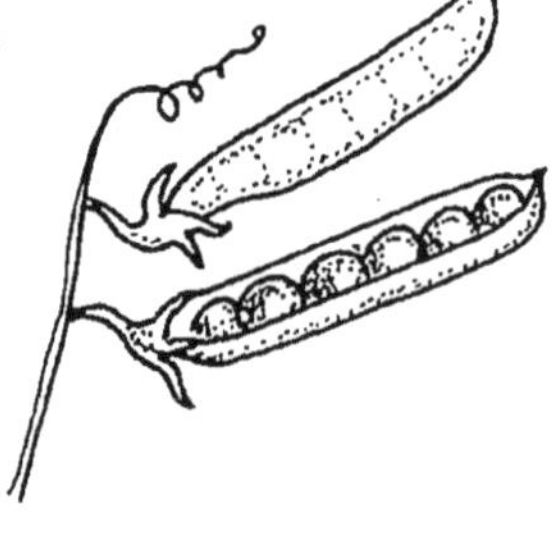

• Pumpkin - insect pollinated.

You will get more fruits if you do some of the pollinating by hand. All squash and the pumpkin family are very susceptible to cross-fertilisation. On cutting the fruit, scrape out the seeds, wash away the casing and dry seeds well.

• Sweet corn - airborne pollinated.

Easily cross-fertilised so keep only one variety for seed. As the pollen is wind distributed, corn is grown in squares, not in rows, for better chance of pollen reaching every one of the "silks". The male flower is formed at the top of the plant, sometimes three metres tall. When the seed cob is ready for pollination, pollen will fall in large amounts from the flower - you can see it all happening if your timing is right.

Every seed in the cob, which is eventually going to be eaten, has a filament or "silk" attached to it. All the silks reach the outside together at the top of the cob. Pollen has to reach every one of those seeds, so we grow sweet corn plants together as much as possible, so that there is a greater chance of this happening. Wind can divert the pollen away from the seeds it is trying to reach if we have only a single row of plants.

For seed saving, select the ear you want, and when the silks are turning from green to brown (which means they have served their purpose and the seeds have been fertilised), cover with a paper bag to prevent further pollen from bees being deposited.

Also, to prevent inbreeding, select seeds from different plants and then dry and mix. Cobs full of seeds take some months to dry. Allow to dry with the husks on (check for bugs first) and when dry, remove the husks, the point of the ear with the unformed seeds on it, and any mis-shaped grains. Shell by hand and make sure it is quite dry before storing.

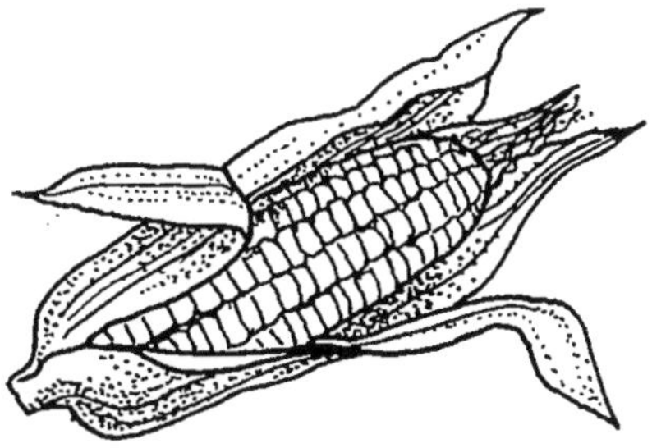

• Tomato - self pollinated.

Tomatoes do not cross-pollinate to a great extent, but a distance between varieties of fifty metres is recommended.

Allow the fruit you have selected to ripen beyond the edible stage, but do not allow to become rotten. Pick and scrape out the seeds, put in a little amount of water and allow to ferment for two to three days in a non-metallic container. Then place in a sieve and wash clean under the cold tap, drying on a piece of paper or cloth until dry.

The reason for the fermentation process is to prevent the spread of bacterial canker disease, which is often perpetuated by the seed.§

Coppicing

A much under-rated method of tree cropping is coppicing. For well over a thousand years, trees have been cut back to stumps, or stools, just above the ground and allowed to re-sprout and form new timber for an ongoing crop.

Some woods in England have been coppiced on a regular basis for at least six hundred years.

In areas where stock may damage new shoots, the trees are cut above the reach of browsing animals. This practice is called pollarding.

Although the tree is severed, the root system survives and is responsible for the dramatic growth that follows. Once the coppice forest is established, the ground does not have to be disturbed.

This means that soil ecology can flourish, undisturbed, through many crops. Also on steep ground, soil erosion is less of a risk. In marginal rainfall areas, the crop does not rely on a consistent annual rainfall.

The greatest potential for coppicing is in the Third World, where, for more than one third of the world's population, the real energy crisis is a daily scramble to find wood to cook meals. Here, more invasive pioneer species, which are both aggressive and fast growing, are adopted.

The Indian neem tree, introduced to the Accra Plains of Ghana early this century, has been providing firewood to towns such as Eumasi ever since. The supply seems to be keeping up with demand as the vigorous and resilient neem is re-seeded by bats that eat its fruit. In the Palibaran area of the Philippines, *Leucaena leucocephala* is yielding twenty cubic metres or more of firewood per hectare per year.

A little closer to home, a well-designed coppice system can supply much of the domestic fuel needs on a suburban block, without reducing the productivity of the rest of the garden. Evergreen legumes which coppice (not all trees coppice), can be grown as understorey to deciduous fruit and nut trees, by supplying them with nitrogen. They also can supply valuable mulch in the form of leaves or if you eat guinea pigs or rabbits, can be converted to meat and manure.

The products of coppice don't stop at fuel and fodder. On our own property we have harvested the coppices of jarrah trees, felled 30 years or so ago. These poles would have been of higher quality if the coppices had been selected and thinned. Most of the *Eucalyptus* species recover quickly from coppice because of their ligno-tubers, which store food below the ground.

Poplars make good joinery wood from coppice and, if left for longer, make good veneer logs. Robinias coppice to make good garden stakes and are coppiced also for gondola poles.

Chestnuts also make good garden stakes and ladder rungs. Ash makes non-splintering tool handles and mulberries will train nicely into pitchforks.

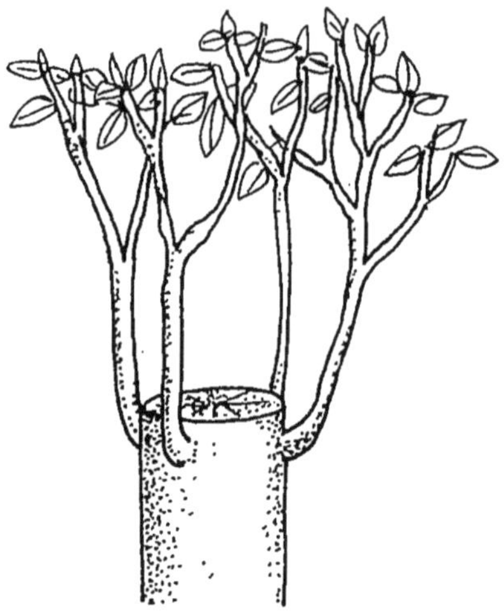

COPPICED TREE

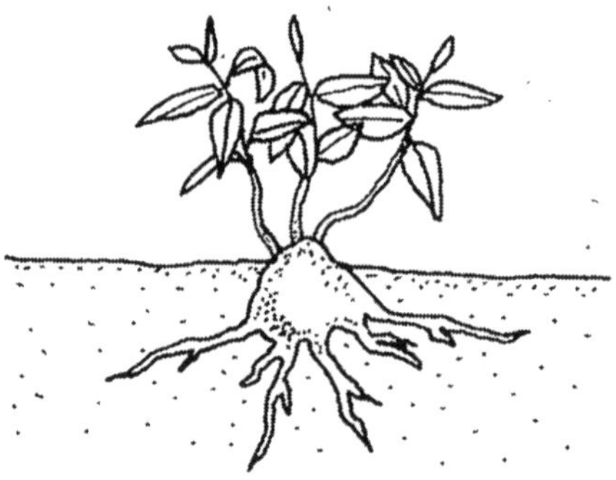

GUM TREE LIGNOTUBER

(R) Canes of coppiced willow.

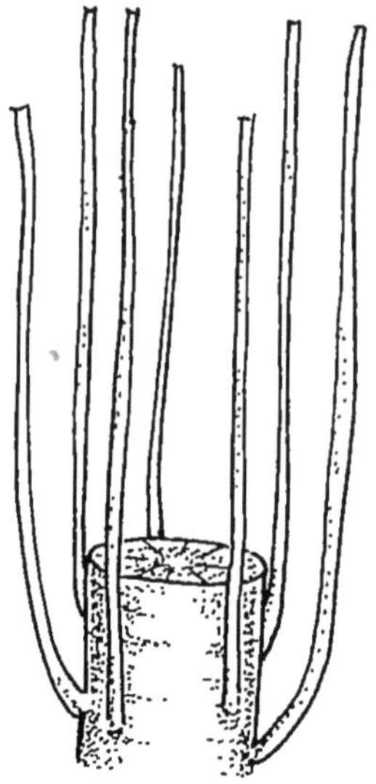

Willows coppice easily. *Salix alba* is called the cricket bat willow for obvious reasons. The osier willow is coppiced to provide long "canes" for making baskets.

Trees are also coppiced for essential oils. *Leptospermum* species are used for tea tree oil. Eucalyptus oil is strongest in foliage after coppice. Of our eucalypts, karri has a 1.2% eucalyptus oil content in leaves, jarrah has only 0.25%, while wandoo and yate have 1.25%.

Many species can be coppiced in the backyard for cuttings for nursery stock; figs, olives and mulberries to mention a few.

Let's not forget the incidental products from our coppice lots either. Products such as fruit, honey, shelter and fodder.

A carefully managed coppice lot will maximise these products. There is much to learn about coppice and much of this will be from trial and observation.§

Aquaculture: the Permaculture Way

Permaculture design allows for multiple and maximum use of one resource. In aquaculture, polyculture is the utilisation of all the levels of the pond or dam.

Many places are limited in the number of freshwater fish species. However, this paucity of fish species is somewhat compensated by native freshwater crayfish species, such as marron, koonacs, gilgies and yabby.

Fish species

Freshwater fish species which have some value in a permaculture system are:

1. The five minnow species which are in direct competition with the introduced mosquito eating gambusia. The minnows are also mosquito larvae feeders and should be stocked in preference to gambusia, which has the habit of nipping the fins of larger fish, thus immobilising them. The western pygmy perch, which prefers warmer temperatures and occupies the same niche as minnows, and gobies (a small fish) may be included in such a system as they are omnivores, eating larvae, zooplankton, and freshwater shrimps as well as aquatic vegetation and detritus.

2. The freshwater cobbler is found throughout streams and dams in the southern part of the country. It is different from the marine species, though it still possesses large, poisonous, dorsal barbs. They would be undesirable in a local aquaculture system as they are mainly carnivorous, eating large crustaceans, worms and molluscs.

3. Carp is another common species. However, this species has the disturbing habit of stirring up the bottom, thus upsetting the habitat for marron or bottom sheltering crustacea for food.

4. The freshwater mussel (found in many southern rivers) is edible and is known to burrow into soft sediments.

5. Rainbow trout and marron are limited in their distribution to particular climatic conditions. Cooler water conditions increase the availability of oxygen which is an important factor in the survival of these animals. Trout and marron should both be compatible in an aquaculture system.

Trout prefer a temperature range of 10-20°C (easily achieved in winter) while marron prefer water temperatures up to 25°C. At all times, it is essential to encourage high oxygen concentrations which can be achieved by having a well-balanced system. Stocking densities of one trout to two m^2 of surface water have been quoted for trout. For marron, stocking densities of 2-4 animals for each m^2 is probably adequate in a permaculture system. The trout feeds primarily on insect life, which should be attracted to the pond by planting selected trees and vines, such as grape, bean and cucumber.

A lack of oxygen within an aquatic system can be distinguished by trout at the water surface, gulping air, and the presence of marron at the edges of the pond. Oxygen must be supplied to the pool immediately by:

1. Pumping air into the pool, thus causing circulation and increasing absorption of oxygen from the atmosphere.
2. Cooling the water down by 5°C to increase oxygen solubility - but be careful as the temperature shock may kill the fish.

Daphnia (zooplankton) in the pond, will indicate that oxygen is present in adequate concentrations, and will also provide good food for trout.

Marron in the backyard are best fed on red manure worms obtainable from compost boxes and worm farms.

All of the above species can be stocked in farm dams or backyard ponds. However, excessive heat and the lack of oxygen will increase mortalities. It is important to:

1. Provide maximum shading possible.
2. Restrict organic pollution. This occurs in ponds from overfeeding and food then decaying, while in dams it occurs by allowing stock residues to enter or be washed in unscreened.

Dams should be deeper than 1.5 m and have sumps or deeper sections to provide refuge for cold water species.

Ponds can vary from plastic wading pools of one to five metres diameter upwards to larger concrete ponds with depths up to two metres in the sump plus shallower shelving areas. Dams should be situated in a very windy spot so that the consequent water disturbance assists in maintaining a suitable oxygen concentration. In ponds, an electric paddle wheel or "water mill" throws up a spray of water, aerating and mixing the pond very quickly.

To stop bird predation, complete cover may be necessary for the pond, or alfoil strips strung across the dam on strings.

Aquatic plants

A large variety of water plants can be used. Some of these include:

1. *Potamogeton pectinatus* (sago pondweed, fennel pondweed) will thrive in saline water up to 5000 parts per million. 24 m^2 of culture produced 35 000 tubers in one growing season. It is good food for ducks and waterfowl - mainly the fruits - but can take over a pond, so maybe some containment is a good idea.

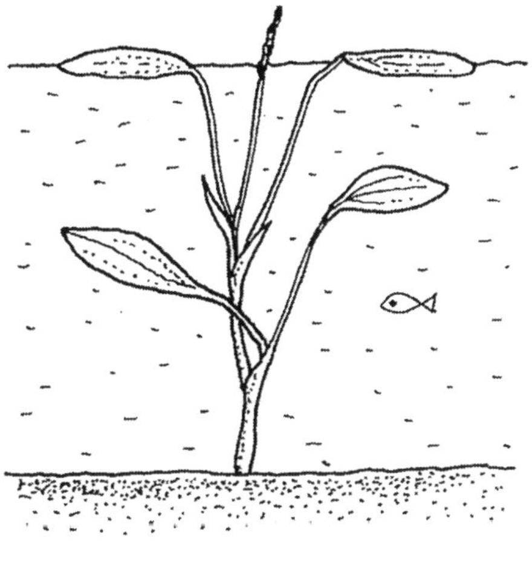

POTAMOGETON

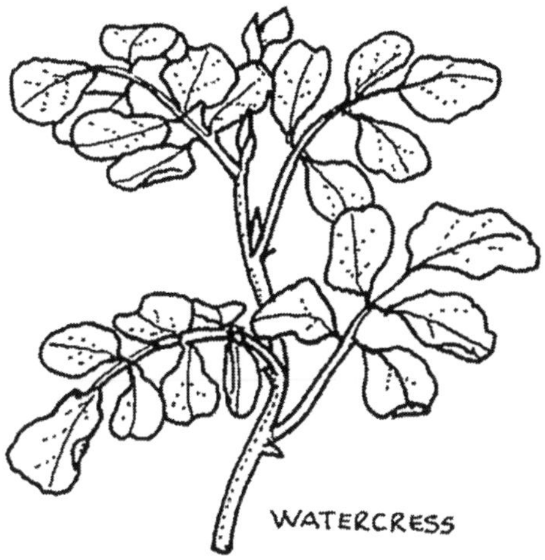

WATERCRESS

2. Scirpus (great bulrush), Baumea (jointed twigrush) and Juncus species. The rhizomes are edible with the possibility of use for flour, and upper stems useful for grazing animals. Good for stabilising banks, bird habitat and aquatic insects. Juncus - controls erosion, good for basketry and seeds, good for fowls. Rushes can be used as filters as they upgrade the water quality of wastewater.
3. Watercress - three to four harvests a year but be careful in case of pollution.
4. Lemna species (duckweed). Harvested at four-day intervals as a nutritious food with 20% protein and 40% carbohydrate. Grow in conjunction with azolla and potamogeton. Good food for ducks, geese and pheasants.
5. Azolla - doubles leaf area in seven days. Grows in still waters and in conjunction with nitrogen fixing bacteria, then harvested as a nitrogen rich fertiliser and mulch. Good duck food.

AZOLLA

6. Typha (cumbungi) - watch that it doesn't take over. Rhizomes are good cattle and pig food. Shoots are good in salads. Used also for basketry.

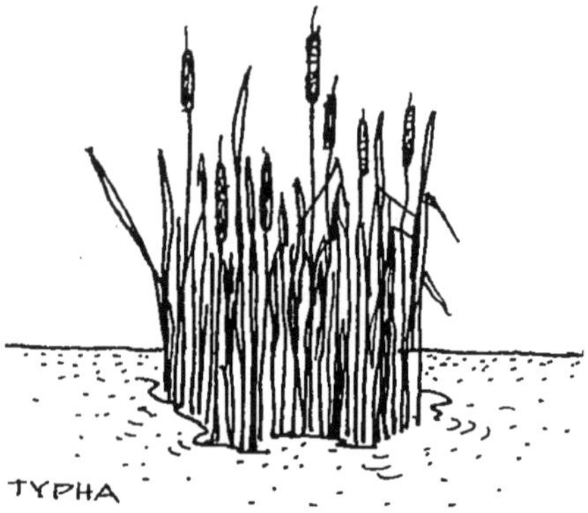

7. Bamboos. Clumping species are often better than running species, although these are contained by water. Bamboos have a large number of uses, including bank stabilising, stock food, shelter and wind breaks.§

Marron in Your Backyard

For those finding it necessary to travel beyond the easily accessible marroning rivers or dams to find a good feed of marron, now is the time to consider growing your own.

Anyone who has a disused swimming pool, fish pond, stream, dam or natural spring on their hobby farm or in their backyard, has the opportunity of introducing marron or yabbies for their own consumption.

Marron are the choice of many aquaculturalists in Australia, due to the large quantity and quality of flesh recoverable from the tail and claws. Also, marron will not burrow holes through dam walls like some other species.

It is interesting to note that the marron is the world's third largest freshwater crayfish.

Identification of species

Yabbies are also extremely good eating but the total flesh recovery is only around 20% of the animal.

Over the years yabbies have built up a reputation for poor taste. This seems to have occurred from the fact that most people catch yabbies in the poor quality water of a farm dam.

Water quality affects the taste of the animals' flesh, and although marron are found in good quality streams of the South West of Western Australia, yabbies seem to be forced to accept the muddy stock-trampled farm dams.

To improve the muddy taste of yabbies it is advisable to place them in good quality, aerated water for a couple of days and the result is an animal comparable in taste to the marron.

Things that affect the survival of marron or yabbies are the quality of the water, food available, predators, oxygen supply and water temperature.

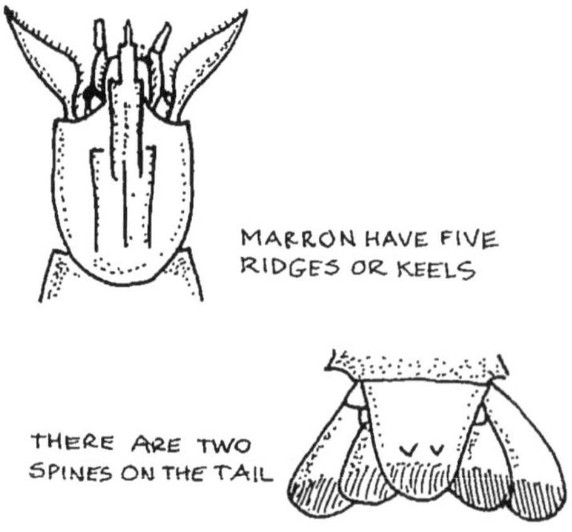

Quality of water

If using water treated with fluorine or chlorine, it is necessary to remove the chlorine before use. This can be done by boiling (which is not feasible with large water requirements), by natural dispersion - allowing the water to settle in a tank which is open to the air for several days - or agitating the water with aeration.

For anyone considering a commercial venture, further details should be obtained of the ideal chemical composition of the water. The backyard aquaculturalist need only ensure that the pH (acidity) level is on the neutral or alkaline side of the scale i.e. 7.0 to 8.0. Although adult marron will survive in salt water the equivalent of half sea water, they will not grow too well and juveniles are not salt tolerant.

Marron and yabbies can be fed on a variety of diets. Well known foods are chook pellets, garden worms, compost, water-soaked grain, potatoes, carrots or even relatively expensive trout pellets. To keep an animal's interest in the food and maintain maximum growth, it has been found that rotation of different feeds will produce better results. When hand-feeding it is unwise to overfeed unless some form of aerating the water is available. The food will eventually break down in water and, in the process, remove the dissolved oxygen which is vital to the animal.

In one suburban backyard, I have seen an excellent automatic feeder in operation. It consisted of a floating 'island' which was made from compost and soil laid on mesh supported by floats. In the soil and compost were garden worms and growing from the soil was a variety of plants and grasses. Onion bags were attached to the corners of the 'island' and dangled to the bottom of the pond as a ladder for the marron to climb to their food.

In another location, I have seen the use of kikuyu grass growing around and into a pond which was then kept trimmed at the edges by very healthy marron. In most cases, experimenting is the only true way to determine if you are providing the correct amount of food. By re-supplying food every one or two days, you can monitor the amount of food consumption. Overfeeding without aeration can wipe out your entire stock; however, underfeeding will lead to cannibalism. Yabbies can tolerate what is considered high levels of poor quality water and in overfeeding conditions will survive much better than marron.

Predators

The worst predator to any commercial farm dam or backyard grower is the two legged predator named *human*. If you intend to grow marron or yabbies and wish to sample the fruits of your labour, make certain your ponds or dams are in an area least accessible to thieves.

Natural predators are varied and include shags, heron, kookaburras, water rats and trout. Some growers provide overhead netting to discourage the birds and have even constructed mini-electric fences to keep out the water rats. Turbid water (milky in appearance) of some clay-based farm dams can offer protection against bird attack; however, water rats appear to have excellent sight even in poor visibility water.

Aeration

In the farm dam situation, it is not economic to artificially aerate, and this is left to the action of the wind blowing across the surface of the water, creating waves and natural aeration. In ponds or commercial locations, aeration is important. Forcing air into the water is available by several means. Airstones attached by tubes to an aquarium-type pump may be sufficient in small ponds. Submersible water fountain pumps which spray the water into the air, or even pumps which have an outlet just above the water and jet the recirculated water back into the pond are also successful.

Construction of a waterfall for recirculation of the water is also an attractive aeration device. Commercial growers will use larger devices such as floating paddle wheels, which are anchored to the middle of the dam and use paddles driven by an electric motor to break up the water. The larger aeration devices may only need to be used three or four times per day depending on their efficiency. Algae blooms (greening of the water) can remove large quantities of oxygen from the water at night, and an automatic timer set to run a pump two or three times at night may be necessary. Yabbies have a reputation for being quite hardy and able to tolerate extremely low levels of oxygen - possibly two or three times less than that of marron.

Water temperature

Marron feed and grow well at 25°C. Higher than this they become stressed and are inclined to become disinterested in living. Lower temperatures reduce their food intake and growth rates. An increase in water temperature during spring triggers the mating mechanism, and the average female will provide around 300 juveniles (fingerlings) usually prior to Christmas.

Marron and Yabbie stocks

Marron are usually sold by age group and those under 12 months are known as nought plus, which is expressed as O+. Between 1 and 2 year olds as 1+ and so on. Marron or yabbie O+ will cost around 50¢ each ordered in lots of 100. Delivery is usually made in a plastic bag containing the marron, water and a 'shot' of pure oxygen before sealing the plastic bag.

When introducing the marron or yabbies to their new location, the entire unopened bag should be placed in the new water and left for half an hour to allow the slow adjustment of water temperatures. The bag can be slit with a knife and the contents released.

Shelter

Old car tyres, lengths of PVC piping, onion bags or the like, all make suitable homes for the animals. It is ideal to provide a ratio of two 'home' or hiding places per animal. Lack of shelter will cause territorial feuds and deaths.§

Hon Matai (Eleocharis dulcis - The Chinese Water Chestnut

Introduction

Although there are many forms of *Eleocharis dulcis* throughout the world, Hon Matai is the one extensively used commercially. The original corms were imported to the USA from southern China and underwent extensive trials. The variety is the only one grown in the USA and is grown as an annual cash crop. It has been successful in the southern states of the USA, and in southern China, and is therefore climatically suited to Australia and many other parts of the world.

As a pond plant, it produces an ornamental reed of up to one metre in length, plus the added bonus of a cash crop at the end of the year. Water chestnuts are crunchy and, when cooked, do not lose their texture. They also contain an antibiotic substance.

How to grow

a) Handle corms and plants carefully - do not damage the shoots.

b) Prepare a sandy-type soil with a good amount of balanced fertiliser or well-rotted manure.

c) Fill a large container or bucket with the prepared soil, dampen, plant the corms with their shoots facing upwards then cover them with 50 mm of soil or heavy sand. (Note: If you have purchased corms sprouting in a perforated black grow bag, do not disturb the plant. Just plant the whole bag and cover with 25 mm of heavy sand.)

d) Keep the plants damp or wet until the shoots are 100 to 200 mm tall, then either top up the container with 75 mm of water or, if the plant is in a pond, lower the container deeper with at least 75 to 150 mm of water cover.

e) Do not allow the plants to dry out.

f) First growth will be runners developing into a one metre tall ornamental reed. Second growth will be the development of corms under the soil. Regular feeding, preferably in the evening, will increase the crop.

g) After 210 frost-free days, lift the container and drain, then stand dry for approximately 1 month or until needed. This gives immature corms time to mature.

h) Harvest the crop. Eat the large corms (approximately 60%) whilst the small corms make good chook food.

i) Corms can be stored in an air-tight container, in the refrigerator, until needed.

j) Retain some corms for replanting next season. The leaves can be used as fodder or mulch.

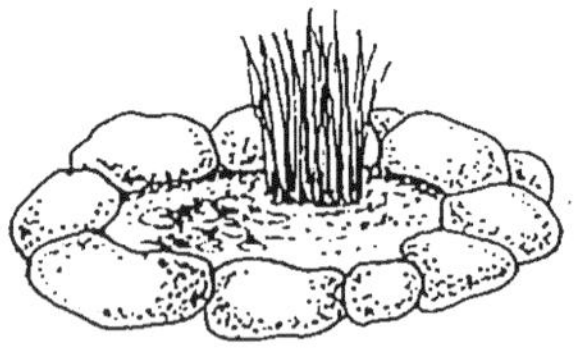

A good guide to planting

Plant anytime throughout spring. They will grow through until next winter. Harvest, eat or store and then grow again. In cooler areas, corms can be started in a pot, in warmer areas they can be planted anytime.§

Gley Ponding

Introduction

Gley ponding is a method mentioned in permaculture literature as a method of sealing ponds and dams.

A gley soil is one in which the drainage is impeded. Water will not drain freely and becomes ponded. It is the result of the breakdown of organic material without the presence of oxygen (anaerobic decomposition). This is directly opposite from humus, which is a product of the breakdown of organic material in the presence of oxygen (aerobic decomposition).

This technique appears to have been used to seal ponds and dams. It offers designers the ability to put a pond/dam in a plan independent of the soil type.

Construction method

1. A depression is dug in the ground. The size can vary from a pond to a dam. It is ideal to have a shallow and a deep end. This enables a diversity of habitats for the aquatic life within the pond. The sides, due to the nature of the gley material, have to be sloping.

Terraces on the sides give an increase of diversity of habitat and allow an escape route if children fall in.

The soil dug out can be used elsewhere - the top soil into garden beds or nearby rockeries and the subsoil to fill any depressions in lawns and paths.

2. A layer of moist lawn cuttings or similar material is spread over the depression's surface.

3. A layer of fresh animal manure is spread over the first layer.

4. A layer of clean, weed-free subsoil or sand is spread over the first two layers. A layer of newspapers is optional between layer two and three. This is mainly an insurance that the first layer does not start to grow, i.e. grass seeds. This is not a serious problem as the water will kill any non-aquatic plants growing in it.

Filling the gley pond or dam

Water is first gently directed into the depression, deep end first. A considerable amount of water is needed to produce gley. It is desirable to have a constant flow of water into the pond until the gley layer is formed.

This is indicated by the presence of air bubbles coming up from the bottom. These air bubbles should contain methane, a gas which is a by-product of anaerobic decomposition.

The exact amount of time required for this process will depend on the size of the pond or dam. During the summer, unless in the shade, it will need topping up every now and then, depending on the temperature.

The overflow from the rainwater tank or gutters could be directed into it via a 'natural' stream. Aquatic plants and animals can be added once the pond or dam starts to hold water.

Gley ponding is great fun. It is a low cost, natural method of sealing ponds.§

Poultry in an Urban Environment

Poultry animals broadly suitable to the confines of suburbia are limited to domestic fowl (chooks), ducks and game fowl (bantams). Since I believe egg-laying to be a more useful and efficient function than rearing only for meat, only egg-laying breeds were sought after. Suitable breeds may therefore include the following (although cross-breeds are also good hardy layers):

- Chooks - Rhode Island Red (brown), New Hampshire Red (brown), Australorp (black), Suffolkshire (grey), Queen, Leghorns (brown or white).
- Ducks - Khaki Campbells (brown), Indian Runners (white), Rouen (green).
- Bantams - Peking Game (feather feet), English Game (no feather on feet).

Eggs vary widely in colour. Chook and bantam eggs range from brown (from brown birds) to off-white or white (Australorps, Queens and White Leg-horns).

Bantam eggs are more pointed than chook eggs, and are about half the size or less. Duck eggs are slightly bigger than chook eggs, with much larger yolks and less white. They have an oily, smooth shell which may easily become dirty or stained. Khaki Campbells lay white eggs. Duck eggs generally do not store as well as chook eggs (about 2-3 weeks versus 6-8 weeks) and old ones have a characteristic odour, sometimes erroneously attributed to all duck eggs.

Chooks have been bred to be consistent layers (and hence poor brooders), producing 250 or more eggs per year, beginning after the first heavy rains of winter, laying daily, thereafter until summer when they, in turn, moult. Bantams lay erratically, with anything from 100-300 eggs per year; due to their broody nature, they go off the lay easily.

Housing arrangements are infinitely variable. There are almost universal local government regulations concerning poultry housing, chiefly concerned with hygiene. Copies may be obtained from your local council offices. Generally, they are only formally upheld following a complaint, but several aspects are obvious. Enclosures should be kept clean, well ventilated, rodent proof, provide night-security against predators (cats and foxes). They should be constructed to provide shelter against rain and prevailing wind; roosts should be available for the fowl at night, suitable dark laying boxes should be provided.

There must be clean water available to the chooks at all times. Traditional poultry-keeping is done on "deep-litter", in a "house and run" or free-range.

Deep litter is analogous to a compost bin. The chooks are kept permanently in a shed, whose floor is covered deeply in sawdust. The addition of manure and constant scratching and turning, produces a rapid bacterial decomposition process (compost) only requiring occasional addition of acid material, like sawdust, and raking of droppings uniformly over the surface. Generally a good system, but not particularly suited to the confined backyard situation.

Similarly, one may attempt to free-range, but the fences are still there and ecological havoc may easily be the result. Thus a compromise is suggested.

A permanent covered shelter with adjoining bird/cat-proof enclosure would allow the animals to be locked up securely, with access to a larger run.

The run may be sectioned off to provide a rotational system of inevitable ecological havoc where the animals are, chook-food growing areas and domestic vegetable growing. Thus, two out of three times use may be made of the ploughing and fertilising functions of chooks, and in the open section of the wire enclosure, straw, droppings and food-scraps can ably supplant the role of a compost tumbler!

It should be noted that, in general, a surprisingly small number of animals can totally remove vegetation from a surprisingly large area if allowed to do so. Rotation is the answer!

I should point out that it would be wise to obtain all birds more or less at the same time when starting a poultry run. This avoids the friction of re-establishing the peck-order repeatedly. Any newcomers have a hard time for several days when introduced.

Food and clean water should be available at all times - this is a trial, since ducks like putting food and sand or mud in their water, while chooks need very clean water. Ducks usually need to have access to water always. They like to wade or swim, or at least dunk their heads and muddy the water, no matter how big or small the container. You can use a large plastic dish in the enclosure, or an old bath with ramp and inset ledge, in the run itself.

This water only remains clean for about half a day, even though duck droppings are rather liquid, unlike chooks. Thus you can have a weekly liquid manure solution to drain on to the garden. If water is not able to be provided, Indian Runner ducks should be sought, as they have much less need for water than Khaki Campbells.

The covered shelter needs roosts and several nesting boxes. The floor should be cement slab, covered in straw. To avoid pecking order squabbles, roosts should be at one height only.

The birds like dark, secluded corners to nest in, generally, although ducks will often lay anywhere (even in water). Bantams and chooks will only lay in a clean site in the straw, free from droppings. Nests should be regularly cleared of eggs to avoid both unwanted brooding, and broken eggs. Apparently, chooks can learn from eating broken eggs to purposely break others for food.

The soiled straw from the chook-house is periodically raked out, and the chooks turn, shred, and add droppings to the straw; their scratching for grain removes weed seeds and mixes in soil, allowing earthworms and bacteria to rapidly decompose the straw-manure mixture. Kept moist in winter, the contents of the enclosure can be turned out as an excellent composted mulch in only a couple of weeks.

I might note that ducks generally compensate for the chook's scratching ability by flattening the ground down again, also splashing water around from their bath. Bantams also scratch, or plough, rather well and the combination of birds does show some advantages.

For your own convenience, the enclosure should be tall enough to enable you to stand upright, and there should be a reasonably large access door. Within this door, a hutch door can be mounted, with a bottom-hinged door to act as a ramp when open. This door needed to be constructed in our case to accommodate the ducks. Entry needs to be flat enough for them to negotiate it and the ramp angle should be slight, perhaps with slats for footholds. This is also the case for access to any elevated bathing vessel.

The door should be tall and wide enough to cater for the extra size of the ducks. Extending out from the enclosure is a large chook-run, a fenced section which can be used to grow chook food, such as sunflowers, alfalfa, borage and wheatgrass, and to allow them to collect fallen fruit. You can place mulberry and fig trees in the pen for summer shade.

Consider extending the chook run temporarily, around the compost heaps, or onto a section of the vegetable patch in preparation for seed planting later - using fences of chicken-wire, varying from one to two metres high.

If flying is a problem the feathers on one wing may be clipped, which soon prevents birds from trying to fly. Of the various breeds, Australorps and New Hampshire Reds are not prone to fly at all.

Chooks and bantams can be handled by the feet, or around the body if tame; ducks, however, have very fragile legs. They should be held by the neck, which is strong, or about the body and wings. You can minimise the use of laying pellets by supplying ample quantities of the three requirements - grains, mash, or kitchen vegetable waste, and greens - with the added supply of grubs, and a calcium supplement (for harder egg-shells).

Grains can be wheat, or any other cereal, sunflowers, alfalfa, tree lucerne, carob pods, or weed seeds for that matter. However, too much wheat can fatten a chook, causing her to go off the lay temporarily.

For their mash, general kitchen vegetable refuse is supplied excluding citrus peels and banana skins and tea or coffee-dregs. Hard vegetables such as pumpkins, carrot-tops, beetroot and parsnips need to be softened by boiling before the chooks can really eat them. Also in this category are fallen fruit - mulberry and figs - helping avoid fruit-fly build up.

Chooks also like uncooked dough and lush greens, in the form of grass, weeds, sunflower leaves and flowerheads, comfrey, borage, cabbages and lettuce. Ducks have difficulty over tough, dry or over-large pieces and need the chooks to shred them first. Grubs are eagerly sought after, especially slaters and earthworms. Snails are devoured by the ducks, and to a lesser extent the chooks. The ducks also delight in pecking at any flies resting on fences. To maintain the quality of the egg-shells, we feed them back, after being blended or crushed, along with shell-grit or cuttlefish.

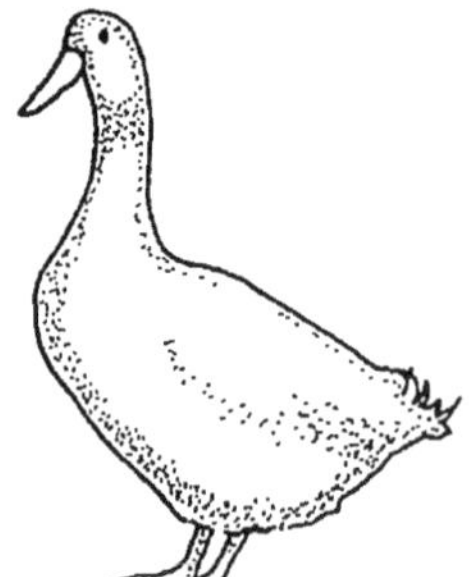

DUCK, CHICKEN, BANTAM EGGS

Grit is also important in the chooks' diet as a digestive "abrasive" in their crops. Being sensible creatures, they can choose how much of anything they want to eat. They don't eat poisonous material such as cape lilac seed or fruit, or potato or tomato plants. Neither do they like strong-smelling plants such as onions or hot capsicums. It is widely stated that chooks need six to eight hens per rooster, that ducks, similarly, need five to six ducks per drake, and also the bantams should be of a ratio of at least two to one.

Apparently if these ratios are not adhered to, the males can bully the females excessively, or conversely, cannot fertilise all eggs when desired. Thus, if only a few eggs per day are desired, it would be generally better to omit keeping any males. Roosters can be rather disruptive annoyances anyway - at all times, but especially in the morning.

Ducks are very sociable and do not like ever to be away from members of their own kind. With regard to noise, the female is the noisier of the pair, emitting loud quacks when hungry, or disturbed; the male however, is relatively inaudible, merely grunting as he breathes in, or hissing when angry. Our chooks and bantams rarely cluck, unless frightened (e.g. by a cat) having pecking-order disputes, or if they see me removing eggs from their nest.

Bantams seem more highly-strung than the chooks, which are rather serious and methodical in their approach to life; ducks, however, are forever playing, in and out of the bath. Their comical antics prove very entertaining. All birds like a regular dust-bath in the sand to remove any ticks or fleas that may be in their feathers. When healthy and laying the comb of either chooks or bantams should be upright and red, not pink or greyish - black areas on the comb are a symptom of stick-flea, a deadly parasite. Ducks are relatively disease-free, not suffering the parasite problems of chooks.§

Bantams - a Useful Addition to a Permaculture System

First we must look at our own requirements like production, appearance of our housing, type of bird and how these things fit into the general environment of our neighbourhood.

We must decide on the area and the number of birds we intend to keep.

Housing is quite easy by using one of the larger commercially designed bird aviaries. On the other hand, you can use a small garden shed two metres by two metres which is a very suitable size, or build your own.

Think carefully about placement. A door facing the east might get strong easterly breezes but protection from rain or the summer sun. Do not put against a solid fence to the sun-side as you will not get enough sun in the winter.

For bantams, one centimetre mesh is needed as small chicks can easily get through 25 mm mesh. Make a pen by building a run of four metres sides, against the shed, with the wire buried 75 mm in the ground.

Take the door off the shed to make the gate. A shed and run of this size will easily take ten to twelve birds.

The birds come next! There are as many different types and sizes of bantams as there are chooks, and we can have several different sorts in this system.

Bantam cockerels are banned in most built up areas, so if we have a broody hen it is best to get some fertile eggs elsewhere. To get a bantam to go off the brood, put her in a cocky cage up off the ground, feed and water well and return her to the flock when she lays her next egg, hopefully in about three days.

Feeding bantams can be to suit your own circumstances. In a situation where you already have other poultry the same food can be used. The easy way out in a totally enclosed run is a galvanised self-serve feeder with pellets, wheat or mash. Mash employs more of their time eating, thus eliminating time for pecking and fighting with each other. Twelve birds will consume around 4-5 kilos per week.

To keep the food fresh you would only need to purchase a maximum of 20 kg at one time. The feeder can be in the shed out of the weather with plenty of clean water in the run, out of direct sunlight. Feed all suitable greens and kitchen waste. Healthy poultry usually do not get diseases, lice or stickfasts. One of the most important pluses with the bantam is the undisputed fact that the food to egg weight production ratio is higher than with the larger birds, but you must remove "broodies" at once to keep up production.§

Ducks in Permaculture

This article is not meant to advise those anticipating a commercial venture but an introduction to ducks for one's own use and pleasure. Ducks are a very useful addition to any permaculture system.

We appear to have three breeds that are most prominent for their general usefulness:

- Khaki Campbell.
- Pekin.
- Muscovy.

The first of these, Khaki Campbell, is the most popular for its laying ability, cleaning up your snail and slug problem in the orchard, vegie garden and so on, spreading their own manure at the same time, or may I say, just pets. In all, this makes the Khaki Campbell the small backyard favourite. It can be housed by simply laying a 200 litre drum on its side, or any other simple structure, so it can lay out of the wind when it wishes. It is common to build elaborate housing, only to find them sleeping under the lemon tree.

The Pekin and Muscovy are the most popular breeds in W.A. for meat birds, their housing requirements being the same as for Khaki Campbells, but their laying ability being much less.

Water

Water is most important to the duck's well-being. Containers need to be deep enough to immerse their heads to drink. These are best placed a minimum of 2 to 3 metres from their feed containers. This keeps down water pollution. Khaki Campbells appear to pollute their water most of all breeds. Ducks like a larger quantity of water to bathe in but this must be kept clean as for drinking containers. I rinse them out every day and scrub out well once a week.

Food

A very useful container can be made for feeding from a section of rain water gutter. Cut gutter in one metre lengths and fit two stop ends.

In the beginning I used chook pellets which I already had available. I then changed over to pullet growers up to four weeks old. From then on, half wheat and half growers. This was recommended by another backyard operator who has been successfully breeding Muscovys, for eating, for many years. The amount I use is worked out on a basis of three-quarters of a kilogram per bird per week, being adjusted according to requirements; that is, as they clean up what they are given.

To supplement this feed with a liberal amount of greens is good practice, feeding an amount they keep cleaned up. My duck enclosures have an added advantage of banna grass and lucerne trees giving shade, windbreak and forage.

Ducks like to forage. I soak the wheat for two days which actually doubles its volume, but I work my quantity on dry weight.

General information

The duck lays at approximately 6 months. Restrict birds at night to enable you to find their eggs in the morning. Most ducks will sit on quite a large number of eggs (about 20). It takes 28 to 29 days to incubate Khaki Campbells and Pekin, but 35 to 36 days for Muscovys, the same as for the goose.

Two things you must never do. Don't leave deep water that young ducklings can get in but not out, or they will exhaust themselves and drown. Don't overstock your property as this causes problems like eating the vegetable patch or possibly, unwanted disease.

Killing birds for meat must be worked out from experience and your own personal preferences.

Just a short guide - Pekin ducks and drakes are almost the same size and weight around 2 kg to 3 kg from 8 - 10 weeks, which is quite a suitable age for the table with the best food conversion ratio for this breed. Muscovys on the other hand vary in weight, the female around 2 kg plus at 18 weeks, whilst the male is around 4 kg at 16 weeks. These are the best times for the table, once again being the maximum food conversion ratio.§

Ducks love to eat snails.

Guinea Fowl

Guinea fowl are long range foragers, although they do return to a pen to be locked up at night. They are deadly on grasshoppers, but they are also deadly on seeds and seedlings, so don't let them near the vegetable garden.

Guinea fowl would have to be the ugliest faced poultry ever seen, but they have mild loving natures, even though the raucous noise made by the birds is enough to scare off the toughest and deafest predator.

The female lays about thirty eggs before they sit to incubate them - once a year around springtime. The chicks are quite small and easily pass through 50 mm chicken wire. They all fly very well, so they should be caged in a pen with a roof, as predators such as owls and hawks like guinea fowl on the menu. Guinea fowl like to roost in trees, so there is no need for a built shelter.

Chickens and guinea fowl don't mix well so keep them apart. Guinea fowl generally harass most other poultry. They will scratch like chooks, but generally won't do as much damage. They tend to wander off to other areas in search of food.

The eggs, if you can find the nest, are pear-shaped, very hard-shelled and about the same size as a pullet's egg. They are easy to cook and taste delicious. Don't try guinea fowl in the towns or suburbs where their noisy calls will annoy your neighbours. They are more suited to small acreage, or larger properties where there is good bush and tree cover.§

Pigeons for Meat - Easycare and Economical

Meat pigeons really are easycare: they are most economical to feed, the squabs are delicious to eat, and the manure is one of the very best you can use.

Feed

You can grow a lot yourself to make a basic 'pigeon mix', by any combination or number of seeds such as corn, sunflowers, peas, fat hen, amaranth, pigeon peas, millet, buckwheat, beans and tagasaste seeds.

They need fine shellgrit and lots of greens too.

A selection of greens should be chopped up each day and scattered in their feed trough, or you can let them out in the garden to help themselves. Fresh water, of course, at all times, plus a shallow, wide dish of water to bathe in.

Housing

This can be basic and inexpensive. Any shed or garage will do, even a small wire lean-to on the end wall of a garage with a small weatherproof shed at one end for the nest boxes would be sufficient. Allow half a metre sq. of space per pair.

Rat-proofing, or regular rat control, is a must, as rats clean up eggs in a flash and upset the birds thoroughly - quite apart from pinching half the food.

Breeding

A pair can be expected to raise ten squabs a year, so a family of four would probably need about five pairs. Incubation takes about eighteen days. Each pair needs two nest boxes - one for the squabs they are rearing and another for the new eggs they are incubating.

Nesting material can be short straw, sawdust, pine needles, or whatever. When the squabs are killed, the whole nest bits, plus valuable manure, can be removed for the garden.

It pays to have similar material covering the shelf floors too, as they are then easier to clean when the manure builds up.

Harvesting

The squabs are killed at about four weeks of age, when they are pretty well fully feathered.

They weigh about a half to three-quarters of a kilogram by then. Don't leave them any longer as they lose weight once the parents stop feeding them and they can sometimes inhibit the parents from breeding again.

Nothing could be easier than killing and plucking squabs - it's a breeze compared to other poultry.

The necks are broken quickly and cleanly, and the feathers plucked dry (none of this lighting up the copper) and they are quick and easy to clean.

Breeds

Good breeds to start with are the Kings or the Carneaux - although crossbreds are often satisfactory. These would complement rabbits and a few chooks.§

Indian Game Poultry

Indian game poultry, properly known as "Cornish Game" are referred to as "the Aberdeen Angus of the poultry world" because the roosters dress out at 4.5 kg or more.

The breed was derived from a Java jungle fowl crossed with an Indian species. Indian Game are an excellent backyard type - attractive and quiet. Being so heavy they don't fly far, and low fences will keep them in. It is easy to breed from full sized birds down to bantams, by constantly using the small ones to breed from.

Indian Game are very attractive birds, with the female having a double laced look in her feathers. They can lay all year round, although they are not noted for their egg laying.

These birds don't usually disturb the neighbours because they have a quiet crow.

An excellent meat cross bird is the Indian Game crossed with a Light Sussex. Jubilee Game, which are predominantly white, are good for plucking.

The White Laced Reds, an American strain, are spectacular to look at.

It is important to maintain a large and varied gene pool for any species, plant or animal, and these birds are no exception. Breeders should keep rare birds such as these to maintain diversity.

There is some gamey flavour in the meat, and the breast meat is white, with more red on the drumsticks.

Although this breed goes broody very easily (eggs must be cleared away every day) they are not good mothers.

Indian game poultry require a deep nest base so that eggs are not broken, as the birds are heavy and clumsy.

Metal houses are better than timber for easier control of fleas and mites. Floor litter can be sawdust, lawn clippings or straw.

These types of poultry can be let out of the pen for a short time each day - for about half an hour before sunset. You will find that they will keep lawn and garden pests down.

Young poultry should be separated from adults and only introduced to the group after they have grown to pullet size. This prevents severe fighting during pecking order establishment.

Feeding

Indian Game can be fed whole grains such as wheat, barley and oats. These can be fed whole, crushed or soaked and sprouted.

Oil seeds, such as sunflower, linseed and canola, are useful additions to the diet.

Plenty of shellgrit is important for general health and providing materials for egg shell development.

Chicks are started off on turkey starter or meal worms. They are normally fed twice a day - early morning and late afternoon. You should not overfeed these poultry as it is not good to keep fat chooks.§

Wormfarming

Wormfarming is the quickest and best way to make compost, as it is four times faster than any other method, and produces an ideal compost which is readily available to the plant and does not heat or burn.

Given the right conditions, the worm population explosion is unbelievable, as is the production of the castings.

Housing or pits

Worms can be grown in anything - from old wash-troughs and baths, to specially constructed wormpits using concrete, or iron sheets for base and sides. The size of the metal constructions depends on the lengths of iron that you can obtain to make an oblong - for example, two metres by one metre.

The secret of the successful pit seems to be to have a shallow reservoir of water across the floor at all times. This means the outlet hole must be slightly higher (10 mm) than the floor, so that any excess water or flooding rain can overflow away, whilst still leaving a shallow reservoir. If using corrugated iron for the floor block up the ends of all the corrugations - both ends, bar the 2 centre ones - with wads of newspaper, or whatever, so that the shallow reservoir of water stays in the majority of the corrugation dips and any excess water overflows out through the two open centre corrugations.

The pits are generally covered with sheets of black plastic; this draws the water up which then percolates back down again. If all the feeding material is put in wet (except the "mill mix"), there is no need to add extra water to the pits. The worms need to be kept moist, not wet, or they will stop breeding.

Food

Anything that has lived and died (though citrus pulp and onions can give the worms indigestion!) is worm food, even your kitchen scraps. Leaf litter is good, and so is any plant material or prunings first fed through the "muncher". Cardboard is well-loved, and grass clippings should be used only in shallow and incomplete layers. Leave some of the pit, for example, around the edges, free so that the worms can escape from any heat generated by the clippings.

Manures are good - both as is, or as liquid manure mixed with shredded newspaper. Commercial chook manure is suspect because there's generally a lot of fly spray in it. Manures can be processed by soaking in a drum of water - any poisons supposedly sink to the bottom in the water, and the manure can then be skimmed from the top and spread on the pits.§

A bran and pollard mixture ("mill mix") can be sprinkled dry on top periodically as an extra feeding for serious wormfarming production.

Another idea for a pit is a plain oblong one for maxi production, as before, but add a centre wall. Here a pit of about 2 m x 3 m, for example, in a wide oblong shape, would have a centre partial wall, and would be filled in a progressive manner. First, a load of manure and vegetable matter would be placed in one corner only, and worms introduced as it cooled down. Thereafter, the next load of foodstuff would be placed in front of this, with the worms moving to the new food as it cooled.

Keep adding loads, but never join the two ends because it's handy to be able to stand in the pit in order to shovel out castings. The original load, now castings, can then be removed with very few worms still in it.

The pits, for maximum production, should be kept at 20° - 25°C all year round. Use grass clippings (in thin and incomplete layers) in winter to give some heat. To insulate in summer use a 5 cm layer of pulped or shredded newspaper which has been soaked in liquid manure. Replenish the paper as it gets used up and pour liquid manure over it.

A worm pit can be any size that suits, from very small, and purely for dealing with kitchen scraps, or larger to deal with offal and so on, when killing one's own animals for meat.

The black plastic pit covers can be made by stapling cardboard sheets between two sheets of black plastic - this gives some weight - and can be held in place by two light ropes strung across the top from hooks on the sides (or those elastic type ropes with end hooks that are used to secure trailer loads).

Worm farming takes all the work out of gardening. Here's how! First select your vegetable garden area. Then start a wormfarm - straight on the ground - at one end. You can use a board to define the outside edge - in which case place all your manures (horse is good) and "munched" plant material along the length of this board.

In the meantime, plant neat rows of your vegetables to fill up the rest of the site. Add your worms to the end heap, and cover with 25 mm pulped newspaper for insulation for summer, or some grass clippings for winter, plus plastic covers. Liquid manure poured on top and/or "mill mix" of pollard and bran sprinkled on regularly will keep the worms well fed and well producing.

Add your new manures and plant materials to the left of the original worm farm line and repeat the temperature regulators and food as described already. The worms will migrate to the new area when they are hungry. You can then plant vegetables directly into the old line which is now perfect castings for growing plants.

Continue in this manner right across the garden. If you really want to hurry things along you can also place manure in the pathways between the lines of vegetables. Add worms, and these will be ready for planting the next crop lines. Eventually, when the whole area is richly composted, you can discard the vegetable lines and get back to companion planting.

Spare worms can be harvested very easily as they come to the surface to feed on the mill mix, or paper/liquid manure. These can be sold, used as a high protein additive for your poultry (particularly guinea fowl) or for your aquaculture residents such as crustaceans or fish, used for freshwater fishing or even for human consumption if 'cleaned out' in a bran mixture for a few days.

Layers of castings can be raked off after the coverings have been removed, as sunlight drives the worms down under. These castings are wonderful for potting mixes or mulches anywhere in the garden. A homemade sieve, using a washing machine motor, can be used to separate larger lots of castings. The worms and undigested matter roll down the slope and away while the castings fall down to the tray underneath. To harvest worm eggs, use layers of cardboard sprinkled with mill mix, and add a few worms. Use layers and layers. In a month, check for eggs.§

No Earthworms! No Garden!

During the time of the Pharaohs in Egypt they thought that the earthworm was a gift from the gods to Egypt alone, and the death penalty was invoked against anyone attempting to take them out of the country.

Since then, all parts of the world that are capable of supporting human and animal life are found to have earthworm populations. Logically, this says that soil that can't support worms can't grow plants for any length of time. Chemical fertilisers make it possible to grow for short periods, but because chemicals do not add humus to the soil, and humus is necessary for all plant life, eventually the ground becomes barren. Add to this the dangers associated with chemicals in the food chain, and organic growing is the only answer.

Because earthworms eat any type of organic matter in large quantities, and turn it into a form that is readily available as plant food, they are a necessity in any growing exercise, and should be treated as such, not seen as a fortuitous accident. Where you have good soil and plenty of organic matter, you will have plenty of worms; where you have plenty of worms and plenty of organic matter you will get good soil, and lots of it. If there are no worms in your garden, then you haven't got a garden! It's not hard to introduce worms to your garden, and it's certainly not expensive. Many people believe that castings are "the finest organic plant stimulator known to humans" and I don't think any person can name one single disadvantage in having worms in your garden beds.§

Earthworms In Your Garden

Of the 3 000 varieties of worms on the earth, four types are common and easy to recognise.

1. The night crawler (*Lumbricus terrestris*), as the name implies, comes to the surface at night (about 10 to 15 cm long).
2. The field worm (*Aporrectodea caliginosa*) is an ideal worm to use in agriculture and orchards (about 10 to 15 cm long).
3. The red worm (*Lumbricus rubellus*) is red in colour and 5 to 10 cm long. These worms are gregarious and breed fast to fill the available bedding and food supply.
4. The manure worm (*Eisenia fetida*), sometimes called the tiger worm because of the distinctive colour rings round their bodies. They, too, are gregarious. They tend to like sloppy, manurey conditions.

People breeding worms usually have a mixture of both red and manure worms, and this does not matter as they basically do the same job:

a. break up organic materials.
b. mix these organic materials into the soil.
c. break up root mats in pasture and thick layers of leaf litter.
d. increase microbial activity in the soil.
e. increase the availability to plants of nutrients in soil and organic matter.
f. improve the crumb structure of soils, and so increase the amount of water that can be held in the soil.
g. allow better penetration of plant roots, oxygen and water into soils.
h. increase crop yields.

The humble earthworm can be harnessed to assist the gardener in reaping better results. It is Nature's own cultivator. Castings are minerally-rich piles of excreta from worms, containing masticated, digested, decayed organic matter of pH 7 (neutral) value. Roots prefer growing in castings which provide balanced plant food.

When mixed with sand, peat moss, or less fertile soil, castings can be used for potting native and fastidious plants, and African violets. To enrich the soil, use 2 parts potting mix to 1 part castings.

Each individual of the earthworm family is both male and female (hermaphrodite), having both eggs and spermatozoa, but it is not self-fertilising.

A capsule of 2-20 worms is produced every seven days, which will hatch in 21 days. The tiny worms are self-sufficient, and will mature in 60-90 days, and are ready to breed when the saddle or clitellum (a thickening about a third back from the mouth end) indicates maturity. They limit their breeding as to space and food.

Food

This means anything organic, and earthworms, like all other living things, need food and lots of it and of suitable quality, if they are to grow and multiply.

Leaves, lawn clippings, weeds, vegetable scraps, compost and old animal manures (chicken litter must be very old). It is not wise to give them too many citrus peelings or onion skins. Food should not decay faster than they can eat it or it turns beds rancid/acid and may smell and also kill the worms. Treat this souring by a gentle sprinkling on, and watering in, of ground limestone or dolomite. Feed little and regularly - particularly wet food. Feed only one part of the bed so the worms can escape if conditions are not suitable; they will migrate if the bed gets too hot or wet.

Beds and Pits

A permanent bed of 150 x 90 cm is recommended as a good size, with built up sides to about 30 cm of concrete, bricks or wood. Bedding, the medium in which worms live and breed, is placed in a dug out pit, approximately 15 cm.

Care must be taken to see that this material is not too hot, so use only well-rotted compost or any other substance. Some of the mixtures that can be used as well as compost are coffee grains, urine, peanut husks, grass clippings, shredded seaweed, vegetables and fruit scraps, torn up cardboard (no coloured ink).

The bedding must be below 27°C after 24 hours, but if above, water well, and check next day. Place worms at a few places on the bed.

Beds having contact with Mother Earth have advantages for worms:

1. Liquid and salts leach away naturally.
2. Escape from heat and food decay is easy.
3. Access to the fine grit for chewing.
4. Bacteria and soil micro-organisms enter and assist.
5. Low cost sides only required for building.

Any container will do for a bed as long as it is at least 25-30 cm deep, such as cut down drums, or an old tub or trough could be used.

Temperature and moisture requirements need to be kept in mind when selecting a site for containers. Earthworms can be bred in boxes with holes in the bottoms and stacked to save space.

As the worms like darkness it is necessary to cover beds with damp bags or carpet underfelt at all times. As well, covers keep moisture in, allow rain to penetrate, stabilise temperature, keep flies away and prevent odours.

Pesticides, weedicides, poisons and chemical fertilisers are bad news for worms. Use less harmful alternatives for all situations - companion planting, pH adjustment, mulching, and legume crops turned in.

Encourage useful insects, such as ladybirds. Use garlic or rhubarb leaf sprays. And practise permaculture!§

Beekeeping

Bees were introduced into Australia from Europe in 1822, but were used for pollination purposes as long ago as 3 000 BC by the Egyptians, and warranted a mention in the Doomsday Book.

A colony of bees can number as many as 50 - 70 000 at the peak of the season in early summer. A swarm is the natural division when the colony becomes too large for the hive.

The colony is made up of three types of bees - all essential for its survival. The queen has an important reproductive function. She can lay up to 1 500 eggs a day in the spring. She also provides a cohesive force in the colony by secreting the queen substance. The queen bee can survive for up to 2 to 3 years.

By far the largest number of bees are the female workers who serve many functions, such as housekeepers, foragers, builders, undertakers and air conditioners to name just a few - not surprisingly they only live up to 6 weeks.

The male drone's only function is to mate with the queen and so ensure the continuation of the colony. Eight days prior to the swarm, the queen pod is laid and fed with Royal Jelly by the workers - so developing ovaries and producing queens. The first queen to hatch destroys all the others to prove her supremacy.

The virgin queen then mates in the air with one (sometimes more) of the drones, who then falls to his death after performing his life function. The drones survive for up to 4 to 5 weeks, and those that are still present in the hive at the end of summer are starved of food by the workers and left to die of exposure, as their time of possible usefulness has passed.

Commercial beekeepers move their hives up to five times each year to catch the best bee fodder as it comes into flower, but it is possible with careful planting to maintain the hives on one site permanently.

The products from the hive include:

- Pollen, which provides the protein element in the bee's diet and is necessary for the development of larvae and bees alike.
- Nectar, which provides the carbohydrate, water and other essential elements in the diet. Nectar is high in water content and is reduced to an optimum of 12-15% by continual manipulation and regurgitation by the workers to make honey.

  A higher level of water content will cause the honey to ferment. The hives would be robbed about 3 times each year giving a nominal yield of 100 kg/year.
- Propolis, which is found by the bees in such trees as poplars, willows, pines and oaks, is used to seal up the hive. It has been found to contain important antiseptic and fungicidal qualities. Pure propolis fetches a high price.
- Beeswax. There are eight wax plates on the undercarriage of the worker bee which are manipulated into the wax cones that are built on a mathematical formula called a "catina". The wax is used in candles, polish, cosmetics, preservatives and lubrication.
- Royal jelly - large export potential, but not fully developed.
- Pollination. Large potential but often not a recognised service for farmers and orchardists.

  Trials have proved that where pollination by bees has taken place the yield can double in fruits such as blackcurrants and strawberries. Other fruits that would benefit include apples, cherries, gooseberries and kiwifruit.

A swarm can be obtained by contacting your local shire, pest controller or the Department of Agriculture. A few shires and towns will not allow beekeeping so it is best to check first.§

Starting Out With Bees

Beekeeping for home use and a little cash can be quite rewarding, as it is a fascinating hobby. However, beginners should not jump into collecting and keeping bees until they are certain they can put up with a little discomfort from the odd sting, and that bees don't worry them.

If you are over that hurdle, then I suggest you initially study as much literature as possible. The State Agriculture Departments have quite a few free handouts and some booklets for sale on bees and hives. There are several books in the library system on the subject, all of which will add to your knowledge on how to handle them.

I also recommend doing a term at night school which usually has a couple of practical sessions at an apiary.

By this time you should know whether you want to continue and how to continue. If this knowledge is gained in autumn and winter you would first prepare your boxes before spring, ready for collecting your first swarms.

By registering your name with the Ag. Dept. it can be given to people with unwanted swarms who ring in to have them removed, and you go and collect the bees in your prepared hive boxes.

Boxes can be made at home but must be to specific dimensions. These are found in the Ag. Dept's literature. The frames are a different kettle of fish. They may be made at home but are quite fiddly. Boxes and frames and wax foundation can be bought at several beekeeping equipment suppliers found in the Yellow Pages.

The first swarms available in spring are generally the best as they are the strongest.

A strong swarm would be about twice the size of a football, but anything from football size up is good.

Some people charge to collect a swarm to help cover time and travel expenses, but often you can get a swarm free or find someone who wants one.

It is a fairly easy process to box a swarm if the swarm is accessible - such as a branch of a tree. But if the branch is high or the swarm is in an awkward place such as a ceiling, cavity wall or tyre well of a car, you can have trouble.

Get as many bees into your box as possible, seal the box and take it home. The unboxed bees will return to the original hive.

There are many permaculture people keeping bees and I feel sure that any of them would be only too willing to give advice or practical help if requested.

Also, for practical experience, most commercial beekeepers would welcome a hand for a day or so. You have now started. Good Luck!§

Rabbit Keeping

Housing is the first basic consideration before obtaining your rabbits. It needs to be:

- Strong - to provide adequate protection against predators such as dogs, foxes, hawks and cats.
- Waterproof - damp conditions lead to health problems and encourage mosquitoes, hence the risk of myxomatosis. The cage can be covered with flywire as extra protection against mosquitoes.
- Shaded - from direct sunlight to avoid heat stress.

Types of housing include conventional structures, underground (to simulate natural conditions) or dual purpose; for example, a cage built over a worm farm whereby the worms can utilise the rabbit droppings.

Feeding is the next area requiring your attention. Water is top priority. Mishaps have occurred through lack of water, particularly to suckling does. Adult rabbits require 300 mL per day. Suckling does require four litres per day.

Food can be a mixture of greens or dry food such as grains, pellets, seeds and carobs. The wild rabbit usually has its litters to coincide with adequate green grass.

Roughage and fibre from carrots, apples, lucerne tree and banana leaves is also important.

Mating is not without its problems. There is no guarantee that a male rabbit will successfully mate the female, or that the female will necessarily co-operate. False pregnancies known as "cloudbursts" do occur. There is also a phenomenon called "reabsorption" which can occur at any stage of the pregnancy, whereby the foetuses are re-absorbed, usually in response to some external stress such as a change of ownership or environmental factors such as drought. Some simple tips about mating include:

- Take the doe to the buck's cage rather than vice versa to avoid a squabble.
- A doe is capable of mating and conceiving within 24 hours of giving birth to a litter, but the recommended rest period is three weeks between birth and re-mating.
- Gestation period is 31 days with some variation e.g. 28-36 days.

- A sign that the birth of a litter is imminent is a nest of fur, pulled from the doe's belly on the day of delivery, appearing in the nest base.
- Working bucks should only service two does per day.
- Separate the buck from the female about to produce a litter to avoid the possibility of the young being eaten by the buck.
- Never keep two mature bucks in the same cage during the mating season - the dominant buck will 'harry' (harass) the secondary buck, preventing it from eating and drinking, and can severely wound it.§

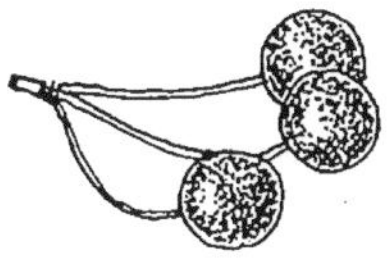

Greens for Rabbits

Practically any grass, herb or weed can be used for rabbit food. My base feed is guinea pig pellets and tagasaste branches and leaves, plus carrot and apple and a carob bean each day. To this, I add a selection of at least six of the following each day. The secret is variety - choose different ones whenever you can.

Greens can include: milk thistles, alfalfa, wild radish or turnip, nasturtium leaves, stems and flowers, borage leaves, stems and flowers, parsley, English dandelion leaves, comfrey leaves, celery herb leaves and stems, all green coprosma leaves and stems, elephant grass, and calendula flowers. Now and then you can include fennel leaves and stems, shallots or chive, N.Z. spinach tips, seven year bean leaves (not beans), lemon balm leaves and stem, marshmallow leaves and stems, fat hen leaves and stem, sweet potato leaves or end runners, and vegetable greens such as chicory, endive, lettuce, cabbage and broccoli.

If you don't have many greens, then include alfalfa hay each day. The health and strong growth of rabbits is due to the vitamins and minerals they obtain from the wide variety of herbs and greens that they get.§

Goats as a Backyard Alternative for Milk

Production Potential

Common types of milking goats include Saanen, Toggenburg, British Alpine and Anglo-Nubian.

Saanen, Alpine and Toggenburgs give very similar yields according to backgrounds and breeding, with Nubians giving a lower yield but a higher butterfat content.

If buying a goat for family milk, it is just as well to buy the breed you fancy as they do become family pets.

It is also far better to have two goats as they keep each other company and are less likely to try and escape from their enclosure.

Housing

Housing must be well ventilated but draught-free. For example, an open front facing away from prevailing winds with ventilation round the top is quite satisfactory.

Otherwise, a garden shed or existing garage or the like can be converted for this use. It is preferable to have a shed around three metres square with two metres headroom, as the herd might increase to a minimum of three from time to time.

Milking and Feed Storage Shed

Milking quarters can be a lean-to with a milking bail on the sun-facing side of the shed.

The alternative is another small shed which will double for milking and feed storage. You could have, for example, several bins with concentrates and hay standing on pallets. This would need to have flow-through ventilation to make it comfortable for milking during summer.

Feeding

Concentrates are necessary to maintain goat condition and produce milk. Roughage is very important to maintain a good balance in a goat's diet.

The following diet is from my own experience but can be adjusted to suit the goat - 1/2 kg goat pellets and 1 kg oats per day and hay in the rack at all times.

To subsidise this I use a mixture of mulberry leaves, brambles, elephant and banna grasses, sugar cane, any willows, loquat, tagasaste, brassicas and many other leaves and branches from an integrated Permaculture system.

Also, a mineral lick is always available. Requirement for hay varies according to the quality and quantity of the green feed available.

Milking and Handling

The ideal is to milk twice a day when the goat is in full production. The udder must be washed and dried before milking. Paper towels are good for this, being cheap and hygienic. Hands must be washed and dried before each milking. It is advisable to use stainless steel or glass containers, with a cover to fit, to milk into.

Plastic can be used but may become impregnated with bacteria after a long period of use.

Milk must be cooled as quickly as possible after milking to keep it fresh. It needs to be strained - two household strainers with a paper tissue in-between is quite alright - and refrigerated immediately.

All equipment used for milking and storage needs rinsing in cold water without delay and then can be washed up or sterilised as you wish.

General Facts

The goat is a browser and eats from 15 cm up from ground level. Contrary to popular belief, it is quite a fussy eater and will not eat food off the ground. It dislikes rain and must not be allowed to get wet as it chills easily and is susceptible to pneumonia.

Goats dislike a sudden change of diet. It must not wear a collar less than 25 mm wide and is better without one. If it must be staked out, do not do so without at least two swivels in the chain and make sure it has easy access to water and shelter from sun and rain.

Keep all plastic materials away from goats as they will not break down in the stomach and will cause digestion stoppages leading to death.

Goats can be a very pleasant part of your lifestyle, fitting into your Permaculture system and increasing your diversity.§

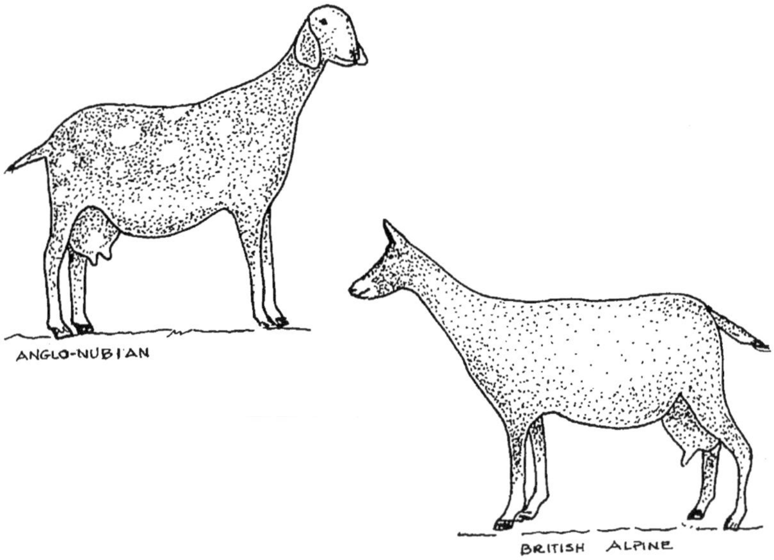

End